MODELLING INTERCEPTION AND TRANSPIRATION AT MONTHLY TIME STEP
INTRODUCING DAILY VARIABILITY THROUGH MARKOV CHAINS

Modelling Interception and Transpiration at Monthly Time Steps
Introducing Daily Variability through Markov Chains

DISSERTATION

Submitted in fulfilment of the requirements of
the Board for Doctorates of Delft University of Technology
and of the Academic Board of the International Institute for Infrastructural,
Hydraulic and Environmental Engineering for the Degree of DOCTOR
to be defended in public
on Monday, 29 April 2002 at 13:30 hours
in Delft, The Netherlands

by

MARIA MARGARETHA DE GROEN

born in Nijmegen, The Netherlands
civiel ingenieur, *Delft University of Technology*

This dissertation has been approved by the promoter
Prof.dr.ir. H.H.G. Savenije   TU Delft/IHE Delft, The Netherlands

Members of the Awarding Committee:
Chairman   Rector Magnificus TU Delft, The Netherlands
Co-chairman   Rector IHE Delft, The Netherlands
Prof.dr. P.A.A. Troch   Wageningen UR, The Netherlands
Prof.dr.-Ing. A. Bárdossy   Stuttgart University, Germany
Prof.dr. M.J. Hall   VU Amsterdam/IHE Delft, The Netherlands
Prof.dr.ir. C. van den Akker   TU Delft, The Netherlands
Prof.dr.ing. W.G.M. Bastiaanssen   ITC Enschede, The Netherlands

The cover shows a detail from a painting of a Zimbabwean rainy season by Mai Brema (mother of Brema). The painting shows, in her words: 'People are ploughing whilst it is dry and with cattle that show they are suffering from not having good grazing. Now they are sowing the crops. After a few weeks the rain comes and the maize grows green. Then the rain never comes back and the maize starts to lose colour. The people continue weeding.' (Ilse Noy, 1992, The art of the Weya women, Baobab Books, Harare, Zimbabwe; design of cover by Peter Stroo)

IHE Delft, The Netherlands   www.ihe.nl

© 2002 Swets & Zeitlinger B.V., Lisse, The Netherlands

All rights reserved. No part of this publication may be reproduced, stored in a retrieval system, or transmitted in any form or by any means, electronic, mechanical, by photocopying, recording or otherwise, without the prior written permission of the publishers.

ISBN 90 5809 378 6

Contents

Abstract

Modelling Interception and Transpiration at Monthly Time Steps - Introducing Daily Variability through Markov Chains

Most of the existing monthly evaporation models simplify the process of evaporation to the extent that they do not make a distinction between interception and transpiration. However, the correct representation of interception and transpiration is very important in water resources assessment, for example to determine the need for supplementary irrigation. The process of interception has a time scale of a day and the process of transpiration has a time scale ranging from a few days to a few months (depending on the ratio of the storage capacity in the unsaturated zone to the potential transpiration). Because these scales differ from the monthly time scale, the modelling of interception and transpiration at monthly time intervals is not straightforward. However, for strategic planning purposes it is sufficient to know this information at a monthly time scale. This creates the problem that daily modelling is necessary to do justice to the combined processes of interception and transpiration, while there often are not sufficient data for daily modelling and while the water manager does not need information at a daily time scale.

The objective of this research has been to improve monthly models through the use of the statistical characteristics of records of daily rainfall at a few locations in the region. This resulted in a new equation for interception. For transpiration it appeared that the existing monthly models could be improved considerably through studying the process at a daily time scale, but that the statistical characteristics of the variability of rainfall within the month do not add much to the accuracy.

The methodology is based on Markov processes at daily time steps, which implies that the probability of occurrence of rainfall on a particular day depends on whether the previous day was a dry day or a rain-day. Markov processes are normally used in stochastic rainfall models. It is shown that the transition probabilities of the occurrence of a rain-day after a rain-day and of the occurrence of a rain-day after a dry day, can be expressed as power functions of the monthly rainfall. This is valid for many locations in the world although Zimbabwe was the foundation for this dissertation. The relationship between monthly rainfall and the transition probability of a rain-day after a rain-day is spatially homogeneous in Zimbabwe. The relationship between monthly rainfall and the probability of a rain-day after a dry day can be estimated from a few daily rainfall series in the region (at a spatial scale of 300 km). Only four parameters are determined to fit the two power functions that each describes a relationship between monthly rainfall and one of the two transition probabilities. Thus, the Markov process offers a key to the variability of rainfall within the month, by use of only a few rainfall stations in the region.

Using the Markov process, the probability density functions for different indicators of rainfall variability within the month are expressed as a function of monthly rainfall: the number of rain-days in the month, the lengths of wet and dry spells, the lengths of the longest wet and dry spells within the month, the number of wet and dry spells in the month, the date of the first day with rainfall occurrence. All these indicators are useful for water resources planning.

For a particular monthly rainfall, the rainfall amounts on rain-days follow an exponential probability density function. The mean rainfall on a rain-day is the scale parameter of this function. This scale parameter does not need to be fitted separately, like is done in stochastic models, but can be expressed as a function of the Markov transition probabilities. Subsequently, the probability density function of rainfall amounts on rain-days and that of rain totals of wet spells is expressed as a function of monthly rainfall.

Combining the probability density function of the number of rain-days in a month and the probability density function of rainfall amounts on rain-days, an equation for monthly interception as a function of monthly rainfall is derived analytically. Not only expressions for the mean and median monthly interception are derived but also expressions for any probability of exceedance. The interception equation uses the daily threshold for interception as an input parameter. This threshold directly relates to land cover and is therefore spatially inhomogeneous but rather constant in time.

In models with a daily time scale, transpiration is potential as long as the soil moisture content is more than a certain value and is constrained proportional to the soil moisture if it is below that value. Existing monthly models did not fully take account of this. The transpiration model of this dissertation is a solution for a constant flux of effective rainfall throughout the month. It is shown that this simple solution, although more advanced than existing monthly models, yields a similar outcome on a monthly basis as a more complex one that simulates the expected variability of rainfall during the month.

It is demonstrated that evaporation models can significantly be improved if a distinction between interception and transpiration is made. Additionally, making use of the expected number of rain-days in the month, improved estimates for monthly potential transpiration are obtained.

For locations where point rainfall agrees to a Markov process, areal averaged daily rainfall also agrees to a Markov process. It is shown that the transition probabilities can be derived from those at the point locations. Therewith an analytical expression is determined for the overestimation of interception by daily water resources models that use areally averaged rainfall as input while correlation between the rain gauge data on a daily basis is insignificant.

The performance of all equations derived has been compared with monthly outcomes of models with daily input. The interception equation is valid in any region where daily rainfall occurrence agrees to a Markov process and where rainfall per rain-day agrees to an exponential distribution. For other climates where the relationship between the monthly rainfall and the mean number of rain-days is known, the interception equation can also be used. The transpiration equation is valid worldwide.

Thus, this dissertation gives improved equations for monthly water resources models that have a transparent link with the physical processes at a daily time scale. The improved monthly modelling of interception, transpiration, potential evaporation and the probability density functions of dry and wet spells is important for strategic planning in water resources management.

M.M. de Groen, 2002

Samenvatting

Het Modelleren van Interceptie en Transpiratie met Maandelijkse Tijdstappen – Introductie van Dagelijkse Variabiliteit met Behulp van Markov-ketens

De meeste bestaande maandelijkse verdampingsmodellen vereenvoudigen het verdampingsproces zodanig dat ze geen onderscheid maken tussen interceptie en transpiratie. Een correcte representatie van interceptie en transpiratie is echter zeer belangrijk om de beschikbaarheid van water en de vraag naar water te bepalen, bijvoorbeeld in geval van irrigatie als aanvulling op regen. Het proces van interceptie heeft een tijdsschaal van een dag en dat van transpiratie heeft een tijdsschaal tussen een aantal dagen en een aantal maanden (afhankelijk van de verhouding tussen de bergingscapaciteit in de onverzadigde zône en de potentiële transpiratie). Omdat beide schalen niet overeenkomen met een maand is de modellering van transpiratie en interceptie met maandelijkse tijdstappen niet vanzelfsprekend. Voor strategische planning van waterbeheer is informatie op maandelijkse basis echter voldoende. Hierdoor onstaat het probleem dat modellering op dagelijkse basis nodig is om het gezamenlijke proces van interceptie en transpiratie recht te doen, terwijl er vaak niet genoeg data beschikbaar zijn voor modellering met dagelijkse tijdstappen en terwijl de waterbeheerder geen informatie op dagelijkse basis nodig heeft.

De doelstelling van dit onderzoek is geweest om maandelijkse modellen te verbeteren door gebruik te maken van de statistische kenmerken van meetseries van dagelijkse regenval van een paar regenstations in de regio. Dit leverde een nieuwe vergelijking voor interceptie op. Voor transpiratie bleek dat de bestaande modellen aanzienlijk konden worden verbeterd door het proces op een dagelijkse basis te bestuderen, maar dat de statistische gegevens van de variabiliteit in regenval binnen de maand weinig toevoegen aan de nauwkeurigheid.

De methode is gebaseerd op Markov-ketens met dagelijkse tijdstappen, wat inhoudt dat de kans op regenval op een bepaalde dag afhangt van of het de dag ervoor geregend heeft. Markov-ketens worden vaak gebruikt in stochastische regenvalmodellen. Er wordt aangetoond dat de twee transitie-kansen van een regendag na een regendag en van een regendag na een droge dag uitgedrukt kunnen worden als eenvoudig machtsfuncties van de maandelijkse regenval. Zimbabwe is gebruikt als basis van dit proefschrift, maar dit geldt voor veel plaatsen in de wereld. De relatie tussen maandelijkse regenval en de kans op een regendag na een regendag verschilt weinig voor verschillende plaatsen in Zimbabwe. De relatie tussen maandelijkse regenval en de kans op een regendag na een droge dag kan geschat worden met behulp van een paar dagelijkse regenvalreeksen (met een afstandsschaal van 300 km). Het is alleen nodig om vier parameters te calibreren voor de twee machtsfuncties die ieder een relatie tussen maandelijkse regenval en een transitie-kans beschrijven. Op die manier biedt het Markov proces inzicht in de variabiliteit van regenval binnen de maand met gebruik van slechts een paar regenvalstations in de regio.

Met gebruik van het Markov proces kunnen de kansverdelingen van verschillende indicatoren van de variabiliteit van regenval binnen de maand uitgedrukt worden als functie van de maandelijkse regenval: het aantal dagen regendagen in de maand, het aantal aaneengesloten dagen zonder regen of met regen, de lengte van de langst aaneengesloten periode zonder of met regen, het aantal perioden met aaneengesloten wel of geen regen, de datum van de dag met de eerste regenval. Al deze indicatoren zijn nuttig voor waterbeheer.

Bij een bepaalde hoeveelheid maandelijkse regenval heeft de hoeveelheid regen op regendagen een exponentiële kansverdeling. De gemiddelde regen op een regendag is de schaalparameter van deze kansverdeling. In stochastische modellen wordt deze schaalparameter apart gecalibreerd, maar in dit geval is dat niet nodig, want de schaalparameter is ook een functie van de Markov transitie-kansen. Vervolgens zijn de kansverdelingen van regenval op regendagen en van totale regenval in aaneengesloten perioden van regendagen uitgedrukt als functies van de maandelijkse regenval.

Door de kansverdeling van het aantal regendagen in een maand te combineren met de kansverdeling van regenvalhoeveelheden op regendagen is een vergelijking afgeleid voor de maandelijkse interceptie als een functie van de maandelijkse regenval. Niet alleen zijn de gemiddelde interceptie en de mediaan afgeleid, maar ook is een vergelijking afgeleid voor iedere kans van overschrijding. De interceptie-vergelijking gebruikt de dagelijkse drempel van interceptie als invoer. Deze drempel is afhankelijk van de lokale begroeiing en daarom ruimtelijk inhomogeen, maar vrij constant in de tijd.

In modellen met dagelijkse tijdstappen wordt verondersteld dat transpiratie constant is, zo lang als de hoeveelheid bodemvocht meer is dan een bepaalde waarde. Als zij minder dan die bepaalde waarde is, is de transpiratie proportioneel aan de hoeveelheid bodemvocht. In bestaande modellen met maandelijkse tijdstap was dit niet volledig verwerkt. Het transpiratiemodel van dit proefschrift is een analytische oplossing waarbij effectieve regenval constant is verondersteld gedurende de maand. Het is aangetoond dat deze simpele oplossing op maandbasis vrijwel hetzelfde resultaat geeft als een complexere oplossing die de meest waarschijnlijke regenvalverdeling gedurende de maand simuleert.

Het proefschrift toont aan dat verdampingsmodellen aanzienlijk verbeterd kunnen worden als een onderscheid gemaakt wordt tussen interceptie en transpiratie. Bovendien, door gebruik van het meest waarschijnlijke aantal regendagen in een maand kan de schatting van de maandsom van potentiële verdamping verbeterd worden.

Voor plaatsen waar het optreden van regendagen op een punt beschreven kan worden door een Markov proces, kan de gebiedsgemiddelde dagelijkse regenval ook beschreven worden met een Markov proces. Er wordt aangetoond dat de transitie-kansen afgeleid kunnen worden uit de transitie-kansen van de punt-lokaties. Op die manier is een uitdrukking afgeleid voor de overschatting van interceptie door het gebruik van gebiedsgemiddelde dagelijkse regenval als invoer in hydrologische modellen met een dagelijkse tijdstap, indien de correlatie in dagelijkse regenval tussen de regenmeters klein is.

iv

De prestatie van alle afgeleide vergelijkingen is vergeleken met maandelijkse totalen van modellen met een dagelijkse invoer. De interceptievergelijking is geldig in alle regio's op aarde waar het optreden van regendagen beschreven kan worden met een Markov proces en waar regenval op regendagen een exponentiële kansverdeling heeft. Voor andere regio's waar de relatie tussen maandelijkse regenval en het aantal regendagen in de maand bekend is, kan zij ook gebruikt worden. De vergelijking voor transpiratie geldt wereldwijd.

Dit proefschrift bevat dus verbeterde vergelijkingen voor hydrologische modellen met een maandelijkse tijdstap die een duidelijke relatie hebben met het fysisch proces met een dagelijkse tijdstap. Het beter modelleren van interceptie, transpiratie, potentiële verdamping en de kansverdelingen van droge en natte aaneengesloten perioden is belangrijk voor de strategische planning in het waterbeheer.

M.M. de Groen, 2002

Preface

This dissertation is not the result of research that followed the original proposal as if it was a recipe.

I went to Zimbabwe in September 1995 to carry out PhD research on the question of whether and to what extent land cover changes had affected downwind rainfall, through changes in the atmospheric water balance. This topic had been initiated by Prof. Hubert Savenije (1995, 1996) and appealed to me. Three IHE MSc researchers and I were welcomed at the Department of Water Resources. In this highly inspiring environment, where large-scale improvements in water resources management were being prepared (revision of water law, institutional reform, new reservoirs), I did research on moisture feedback. PhD and MSc researchers from ITC and Wageningen University arrived and fed me their ideas on crop yield forecasting, remote sensing and groundwater modelling. I received support from everyone and developed new concepts (De Groen & Savenije, 1996, 1998, 1999), but further quantification foundered on the limited availability of data and the fact that I was a novice in climate modelling. Most of the reservations expressed in the dissertation proposal became a reality.

Back in Delft in mid-1999, I was still struggling with the fact that long daily time series in digital format were only available for a few locations. Monthly time series of rainfall were available for some 100 locations and these offered an insight into the spatial distribution of rainfall over the country. However, for moisture feedback processes it is interception that is very important and interception has a time scale of a day. In the meantime, as a lecturer at IHE, I had been the mentor of several MSc researchers who also had only monthly data available and who needed estimates of transpiration and interception (or rather, effective rainfall). They needed simple methods that could easily be modelled in spreadsheets, for lumped approaches, or in GIS, for distributed approaches. (See Section 1.4.)

In Zimbabwe it is common knowledge that there is persistence in the occurrence of rain-days and in the occurrence of dry days. The farmer may not call it this way, but he/she has adjusted his/her practices to persisting dry and wet spells. Having taken notice of this common knowledge and with my problem in mind, I plotted the probability of a rain-day after 1, 2, 3...10 consecutive dry days and after 1, 2, 3...10 rain-days for the three locations for which I had daily data available and for different classes of monthly rainfall (see Figures 3.1 to 3.3). To my surprise, the length of the preceding dry or wet spell did not change these probabilities!

Confident that this could offer a method to better estimate interception and transpiration from monthly data and confident that somebody would have developed such a method, I showed the graphs to Prof. Mike Hall. 'These are Markov chains. Gabriel & Neumann, 1962, see my lecture note on stochastic modelling.' That was a remark which was embarrassing, but very useful to me.

However, it was not merely a matter of finding someone else's solution to my problem, but the start of a completely new research project. The result of this research is in front of you. It is not what was originally planned, but it is most probably of more direct relevance to water resources management.

List of Symbols

Some symbols of models by other authors are not in this list but locally defined in the section where they are used. Also, symbols used in boxes or footnotes only are not repeated here. The parameters which are used in Chapter 10 for potential evaporation of Penman, Penman-Monteith and FAO Penman-Monteith, are explained in Appendix C.

Symbol	Description	Dimension	Unit
β	mean rainfall on a rain-day and scale parameter of exponential distribution	L/T	mm/day
Δ	difference	depends on specification	
γ	time scale for transpiration S_b/T_{pot}	T	days
γ^o	dimensionless time scale for transpiration: ratio between γ and days in month	-	-
λ	parameter in Poisson distribution for number of spells which equals the expected number of dry spells/pairs of spells per month	-/T	1/month
μ_x	mean of x	dimension of x	unit of x
σ_x	standard deviation of x	dimension of x	unit of x
a	fraction of land surface covered by vegetation	-	-
A	intercept of relation between monthly effective rainfall and monthly transpiration	L/T	mm/month
B	slope of relation between monthly effective rainfall and monthly transpiration	-	-
C	difference between transition probabilities. $p_{11} - p_{01}$	-	-
$C_v(x)$	coefficient of variation of x; $C_v(x) = \dfrac{\sigma_x}{\mu_x}$		-
D	daily interception threshold	L/T	mm/day
D_v	interception capacity of vegetation	L/T	mm/day
E	total evaporation (= interception + transpiration)	L/T	see subscript
E_{Gm}	total monthly evaporation according to models of this dissertation (De Groen)	L/T	mm/month
E_{pan}	pan evaporation	L/T	see subscript
E_{pot}	potential evaporation	L/T	see subscript
$E_{pot,s}$	potential soil evaporation	L/T	mm/day
E_{TMm}	total monthly evaporation according to Thornthwaite Mather model	L/T	mm/month
$E(x)$	expected x	dimension of x	unit of x

Symbol	Description	Dimension	Unit
$f(x)$	value of probability density function for x	-	-
$F(x)$	value of cumulative probability density function for x	-	-
F	probability of non-exceedance	-	-
G	groundwater storage	L	mm
I	interception	L/T	see subscript
I_{act}	actual interception per rain-day (only used in chapter 10, otherwise just I is used.)	L/T	mm/day
I_{pot}	energy potential of interception	L/T	mm/day
$M(x)$	median value of x; $F(x) = 0.5$	dimension of x	unit of x
n	number of days	T	days
n_{dry}	dry spell duration	T	days
$n_{dry,max}$	duration of longest dry spell in the month	T	days
$n_{portiondry}$	duration of portion of dry spell that is within the month, for a dry spell that crosses one of the boundaries of the month.	T	days
n_m	days per month	-	days/month
n_r	rain-days for given duration	-	days/month
n_{wet}	wet spell duration	T	days
$n_{wet,max}$	duration of longest wet spell in the month	T	days
$n_{portionwet}$	duration of portion of wet spell that is within the month, for a wet spell that crosses one of the boundaries of the month.	T	days
N_{dry}	the number of dry spells in a month	1/T	1/month
N_{wet}	the number of wet spells in a month	1/T	1/month
N_{pairs}	the number of pairs of a dry and a wet spell in a month	1/T	1/month
p	probability of occurrence of a rain-day	-	-
p_{00}	transition probability of occurrence of a dry day after a dry day	-	-
p_{01}	transition probability of occurrence of a rain-day after a dry day	-	-
p_{10}	transition probability of occurrence of a dry day after a rain-day	-	-
p_{11}	transition probability of occurrence of a rain-day after a rain-day	-	-
p_{only1}	probability that only one of two rainfall stations records a rain-day	-	-
$P(x \mid y)$	probability of x given condition y	-	-
P	rainfall	L/T	see subscript
P_{eff}	effective rainfall	L/T	see subscript
q	constant in $p_{01} = q(P_m)^r$	$(T/L)^r$	$(month/mm)^r$
r	power in $p_{01} = q(P_m)^r$	-	-
u	constant in $p_{11} = u(P_m)^v$	$(T/L)^v$	$(month/mm)^v$
v	power in $p_{11} = u(P_m)^v$	-	-

Symbol	Description	Dimension	Unit
S	available soil moisture content	L	mm
S_b	available soil moisture content at the boundary between moisture constrained transpiration and potential transpiration	L	mm
S_{end}	available soil moisture at end of month	L	mm
S_{max}	maximum available soil moisture for certain soil type and crop	L	mm
S_{start}	available soil moisture at start of month	L	mm
$S_{start,dry}$	soil moisture content at start of dry spell	L	mm
$S_{start,wet}$	soil moisture content at start of wet spell	L	mm
t	time	T	days
t_{11}	recurrence time: if two rain-days succeed each other without a dry day in between, the recurrence time is one day	T	days
Δt	time step	T	see subscript
T	transpiration	L/T	see subscript
$T_{max,m}$	maximum monthly transpiration, given certain initial and soil conditions	L/T	mm/month
T_{pot}	potential transpiration per day	L/T	mm/day
$T_{pot,dryleaf}$	potential dry leaf transpiration per day	L/T	mm/day
W_i	'available moisture'; sum of rain total of month and of available soil moisture at start of month, used by other authors	L	mm

Subscripts

bul	Bulawayo, Goetz meteorological station		
har	Harare, airport meteorological station		
mas	Masvingo, meteorological station		
d	during day, incl. dry days	L/T	mm/day
dry	during dry spell	L/T	mm/dry spell
i	counter		
int	property applicable on rainfall series that are derived by interpolation of the records from two rain stations		
m	during month	L/T	mm/month
n	during nth day	L/T	mm/day
r	during rain-day	L/T	mm/day
t	during time step t	L/T	
wet	during wet spell	L/T	mm/wet spell

1 Introduction

1.1 General

In Southern Africa, time series of monthly rainfall and runoff are more widely available than daily data. Monthly data are generally appropriate for water resources assessment studies. However, in models that use monthly time steps, certain fluxes between components of the water resources system (interception, transpiration, recharge, surface runoff etc.) are governed by daily processes. The question is how the statistical characteristics of daily processes should be used to achieve the appropriate aggregation of these processes into monthly values.

Most hydrological models using monthly time steps do not make a distinction between interception and other forms of evaporation, but prefer to combine all evaporation processes in what is generally called evapotranspiration. However, such distinctions are important. To determine supplementary irrigation requirements in models for water resources planning, one needs to determine effective rainfall, which is rainfall after the subtraction of interception. Also, in future decision making on water resources, transpiration should be accounted for as a productive water flow, so-called 'green water'. Rainfed agriculture is still the main producer of the world's food and solutions to the problem of feeding the world's population will need to come through making better use of rainfall (Falkenmark, 1999; Savenije, 2000). In Zimbabwe, the country for which data are used throughout this dissertation, more than 80% of the population is dependent on agriculture to generate their income, with 80% of these people working in small-scale agriculture (Kupfuma et al., 1992). Commercial agriculture, partly irrigated, produces about 15% of Zimbabwe's gross domestic product and 50% of its exports (Muir, 1994).

Additionally, most global climate models use monthly time steps. It is important to distinguish interception in these General Circulation Models, because it is the direct feedback of water to the atmosphere (Zeng et al., 2000). More important on the local scale is the fact that interception is a latent form of energy during the day after the rainfall occurred. When convective storms take place, the effect of interception on the energy balance triggers rainfall on the next day (Taylor et al., 1997). Transpiration on the other hand is subject to soil moisture conditions, and thus has a time scale ranging from a few days to several months. It affects rainfall on that time scale, thereby providing feedback to the soil moisture, thus causing persistence in rainfall occurrence (Eltahir, 1998; Zeng & Eltahir, 1998). This feedback on two different time scales illustrates the importance of distinguishing between interception and transpiration in climate models. Conversely, this dissertation shows that knowledge on the persistence in daily rainfall occurrence is useful in determining monthly interception.

If it is possible to estimate the relationships between monthly sums of fluxes and stocks of water resources that truly reflect the variability in the daily processes, then this would make the large effort of collecting and analysis of daily data less urgent. In general, at the major meteorological stations (airports) good daily rainfall records are available, statistical characteristics of which can be used to support modelling on a monthly scale. This dissertation thus downscales monthly data by relating monthly rainfall to the statistical characteristics at a daily scale. The knowledge about daily processes is mathematically synthesised to relationships at the monthly scale, which is conceptually clear but limited by the constraints of mathematical tractability (see Blöschl & Sivapalan, 1995).

Because of the prevailing opinion that efforts should be made to change from monthly to at least daily models, recent developments in the equations underlying monthly models have been rare (see references in Section 6.7 and Chapter 9).

1.2   Problem description

The central question is how to model fluxes between stocks in the hydrological cycle, making use of time series with monthly time steps, while catering for processes that are characterised by a much smaller (daily) time scale.

This will be done by asking the following subquestions:
a.   Which relationships describe monthly fluxes as the aggregated effect of daily processes?
b.   How can one derive the parameters of such relations?

1.3   Objective

The improvement of equations for interception and transpiration in monthly water balance models by establishing the relationship between monthly rainfall and the statistical characteristics of daily rainfall.

1.4   The case of Zimbabwe, as an illustration of a generic problem

Changes in water demand (a new irrigation area or the extension of a town), storage capacity (a reservoir) or water use efficiency (water demand management measures) have far-reaching consequences for the availability of water for alternative uses. Such changes can be stimulated by policies or by investment. To be able to choose between alternative strategies, the consequences of these interventions need to be assessed; they influence the whole catchment area and should be considered in a large time frame. Apart from hydrological uncertainties, socio-economic uncertainties need to be contemplated. For these reasons, water resources system models are generally simulation models, and they are run under various socio-economic conditions using long input time series to simulate the possible hydrological spatial and temporal variability. To evaluate different water resources planning strategies for river basins larger than $\sim 0.5 * 10^4$ (km)^2, simulations at monthly time steps are generally adequate.

The methods described in this dissertation are applicable in many climates in the world (see Section 3.8), but data from Zimbabwe are used as the basis. The problem description was inspired by MSc studies that were concerned with topical issues in water resources assessment and planning in Southern Africa. These MSc studies are illustrative of the problems that water resources planning addresses and the role of estimates of interception and transpiration. Makurira (1997) conducted an integrated feasibility study to determine from where Zimbabwe's capital Harare and its neighbouring town Chitungwiza could best withdraw water. In years with low rainfall, the crop yield from both irrigated and rainfed agriculture is low, making the economy very sensitive to hydrological variability. Makurira addressed the question of how different strategies of water withdrawal for drinking water production restrict developments in agriculture and mining, and how robust different strategies are under different economic and hydrological scenarios.[1] Matola (1998) carried out a similar study, trying to ascertain the added economic benefits from a new reservoir and from different demand practices (cropping patterns, irrigation practices).[2] Both Makurira and Matola needed effective rainfall estimates to determine demand from the reservoirs and used crop yield models to simulate agricultural production as a function of water availability. Nyagwambo (1998) studied the financial feasibility of reallocating water from irrigation in the dry season to supplementary irrigation in the rainy season. Much economic benefit is obtained when dry spells in the rainy season, which cause constrained transpiration, are bridged through supplementary irrigation.[3] Nhidza (1999) looked at the impact of large-scale forestry on water resources management, through its effect on evaporation.[4] Seyam (1999) and Adane (2000) studied the spatial variability of water scarcity.[5] As effective rainfall is the most important source of water for food production, estimates of interception and transpiration are important parameters for water scarcity. Better estimates of the spatial variability of water scarcity can serve as preparation for international negotiations (Savenije & Van der Zaag, 2000).

The examples above illustrate the need in Southern Africa for methods to determine transpiration and interception. For similar reasons, such methods are useful worldwide. All the studies mentioned used monthly models, because daily data were not readily available. Another MSc researcher, Mudege (1999) studied the feasibility of different supplementary irrigation practices on a local scale, using daily data. The daily variability of rainfall has an important influence on the estimated crop yields and thus on evaluating the alternatives. This confirms the need to improve monthly models or to switch to daily models, with the latter (becoming) the prevailing opinion. The author of this dissertation believes that monthly models are more practicable than daily models, provided they are improved. This greater practicability is related to the history of measuring and modelling water resources. The example of Zimbabwe is further used to illustrate the arguments for monthly modelling.

---

[1] The strategies considered withdrawals from the catchment of the Manyame and Kunzwi; resp. $3.8*10^4$ (km)^2 & $0.3*10^4$ (km)^2.

[2] Umbeluzi basin in Swaziland and Mozambique $0.6*10^4$(km)^2, thus not in Zimbabwe.

[3] Mupfure catchment; $1.2*10^4$ (km)^2 (see also MSc studies by Shoniwa, 1996; Gbedzi, 1996; Mare, 1998 & Musariri, 1998).

[4] Odzani catchment; $0.2*10^4$ (km)^2.

[5] Zambezi catchment; $1.3*10^6$ (km)^2.

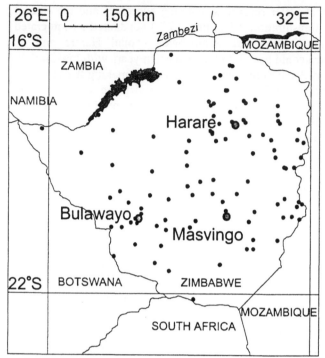

Figure 1.1   Map of Zimbabwe. Data from rain stations in Harare, Masvingo and Bulawayo are used throughout this dissertation. The other dots show the 100 rain stations for which only monthly data were readily available.

In its national water resources assessment, Zimbabwe has so far concentrated on surface water (Ministry of Water Resources and Development of Zimbabwe, 1984, Department of Water Development / Water Resources Management Strategy, 1998). Across the country, more than 7,500 dams store water for irrigation (Rukuni & Makadho, 1994). With dams therefore the most important water resources management tool, surface water is considered to be the prime water resource to be managed.

As a result, the Department of Water Resources of Zimbabwe has spent much effort on developing good rainfall runoff models. For years, the Department has used the Pitman model (Pitman, 1973; see also Hughes, 1995, 1997) with a monthly time step as the standard to extend the monthly runoff records to the generally longer and more complete monthly rainfall records. These extended runoff records are used in feasibility studies for dams, i.e. water rights, and to determine rule curves for reservoirs.

Not all studies are based on monthly data. Several subcatchment studies have used hydrological rainfall runoff models to analyse the daily variability of runoff (e.g. Lørup et al., 1998; Bullock, 1992b). The Department, with the help of the Swedish institute SMHI, has digitised since 1998 many historical records of daily rainfall, available from the Meteorological Department. These data have been used in the HBV model (Lindström et al., 1997), which has a daily time step and which is intended to give better estimates of daily and monthly runoff in relation to rainfall. The developers of this model stress the conceptual form of the model, but the conceptual parameters do have a physical connotation. For example, initial losses and recharge are simulated, although only rainfall and runoff data are used for calibration.

On the scale of Zimbabwe's river basins, water resources planning models with a monthly time scale are appropriate. The variability in daily flow is generally not very important for water resources planning, because average residence times in the ground and in surface water reservoirs are usually considerably longer than a month. To obtain an idea of the consequences of planning strategies, they are modelled for different hydrological and socio-economic scenarios over periods of several decades, a process which would be too cumbersome if done with daily time steps. The daily time steps may improve the hydrological scenarios, but this improvement is small in comparison to the uncertainties in the socio-economic scenarios.

In Zimbabwe, spatial correlation of daily rainfall data is low and the density of the rain gauge network is highly variable across the country (see Figure 1.1). Around Harare the density of the network is comparatively high, because there are many commercial farms. Nevertheless, the Pearson correlation between official meteorological stations in this area is only 0.2 - 0.5 on a daily basis (see Figures 1.2 and 1.3). This is not surprising, because convection accounts for perhaps 90% of the total rainfall in Zimbabwe. A mature thunderstorm has dimensions in the order of 5 to 10 km and a duration of 3 to 4 hours (Torrance, 1981). In the final stages a large thunderstorm may draw on air from 15 to 20 km around, an area in which there is no cloud development. The correlation between monthly data from rain stations at a distance of 20 to 50 km distance is in the order of 0.5 - 0.8 (see Figure 1.3). Therefore, the rain gauge network is not adequate to provide good estimates of daily rainfall on a scale smaller than the distances of the network. However, the current network is fairly good for the spatial distribution of monthly estimates. (Chapter 11 will discuss in more detail the usefulness of interpolation on a daily basis.)

Thus, the interest in daily variability of runoff is small and the spatial correlation between daily rainfall records is low. For both reasons the analysts aim to correctly represent the monthly values in the calibration and verification of daily rainfall runoff models (e.g. Refsgaard & Knudsen, 1996). The criteria for reproducing the daily variability are less strict.

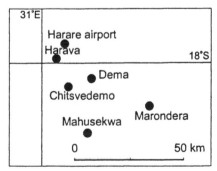

Figure 1.2   Map of the rainfall stations used in Figure 1.3.

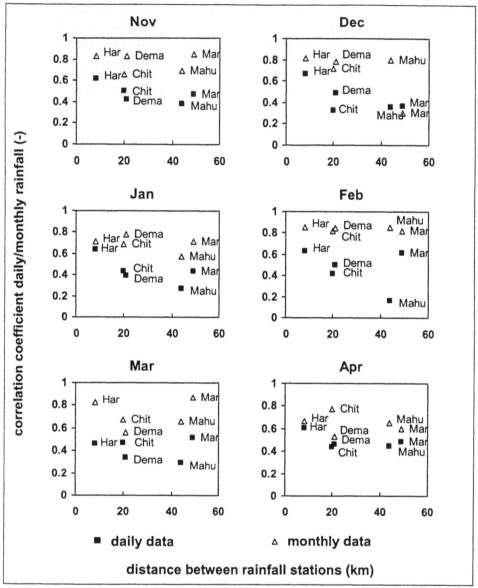

Figure 1.3   Pearson correlation of daily and a monthly rain totals in relation to the distance to the reference station, Harare airport (1959/1960 - 1993/1994). Chit = Chitsvedemo, Dema = Dema, Mah = Mahusekwa, Mar = Marondera, Har = Harava.

This does not mean that daily rainfall runoff models are not important. To design spillways for local and regional catchments areas, flood hydrographs that have a short time scale are necessary. In water resources assessment, daily rainfall runoff models have a more accurate differentiation between the surface runoff and the baseflow components, resulting in better estimates of the aggregated monthly runoff. Other fluxes that are important for water resources assessment, such as interception and transpiration, are also highly dependent on the variability of daily rainfall and are therefore usually simulated using daily rainfall records. All farmers base their decisions directly on the daily variability of rainfall, and the lengths of dry and wet spells are the main factors in determining crop yields in rainfed agriculture (e.g. De Bie, 2000).

It is without question that a good daily rain database is an advantage. However, it requires a rather large investment to prioritise the modelling with daily data, also as it implies first digitising and analysing these data, not to mention the investment in and maintenance of a denser rain gauge network. Relatively new techniques, such as radar and Cold Cloud Duration (Rugege, 2001), can assist in better estimating the spatial distribution of daily rainfall. However, these techniques cannot be used with historical data, from times when no such measurements were taken. Additionally, as mentioned, with water resources system models that are used for planning the interest is not in the daily time scale itself. In such models, the importance lies in obtaining a correct picture of the spatial and time variability of water in the different stocks in the catchment. It is therefore more efficient to put effort into the use of a long time series of monthly records, than into a shorter time series using daily records.

Although the density is highly variable, in comparison to some other African countries Zimbabwe has quite a dense and well-maintained network of meteorological and runoff stations. The daily modelling (HBV model) which is currently being put into practice at the Department of Water Development may be viable for parts of Zimbabwe. Yet it is far less appropriate in other parts of (Southern) Africa, which often have a shorter history of data collection. If the use of monthly data offers a good alternative for planning purposes, it may be worth investing in a denser network that collects data on a monthly basis, rather than a less dense but daily-gauged network of rain stations.

This study investigates how equations for interception and transpiration as a function of monthly rainfall could be improved, so that the monthly aggregates better agree with the aggregated fluxes of daily time steps. Remote Sensing techniques are very effective tools in determining the spatial variability in evaporation (Bastiaanssen, Menenti et al., 1998; Bastiaanssen, Pelgrum et al. 1998; for Zimbabwe, see Lupankwa, 1996; Wolski, 1999; Merka, 2000). However, the processing of such data still requires a considerable amount of time. The methods can only be applied in cloudless periods and as a result it is most often dry season estimates which are made, which is less relevant to rainfed agriculture. Additionally, as with Cold Cloud Duration techniques, this method cannot be used for historical data, even if sufficient money and time were available. Nevertheless, remote sensing is a very useful tool for determining the model parameters in hydrological models, not only to directly quantify the evaporative flux, but also to determine other physical characteristics ('hydrotopes') which are important in hydrological modelling (see Wolski, 1999).

It needs to be stressed that the water resources assessment that this study pursues is used for integrated water resource system models. The focus is on obtaining better monthly relationships for interception and transpiration, which in follow-up research can contribute to a better estimate of other water fluxes, such as recharge or surface runoff.

1.5   A guide through the dissertation and its innovations

The preceding sections have explained that the main objective of this dissertation is to improve monthly water balance models through the introduction of the statistical properties of daily rainfall. The Chapter has defined the importance of monthly water balance models and the need to distinguish between interception and transpiration. Figure 1.4 gives an overview of the approach and layout of the rest of this dissertation.

In Chapter 2 a general introduction is given to stochastic models. Various models are discussed with regard to their suitability to derive analytically equations for monthly water fluxes. The Markov model is the one most suitable. Previous research on Markov-based daily rainfall models is discussed.

In Chapter 3 the theory of the Markov process is described. It is shown that the Markov property is valid for Zimbabwe and that the relationship between monthly rainfall and either of the two transition probabilities can be described by a power or a logistic function. Empirically, power functions between monthly rainfall and transition probabilities are determined for the airports of Harare, Masvingo, Bulawayo and, with less detail, for other locations in Zimbabwe and the world ($p_{01} = q\,P_m^r$, $p_{11} = u\,P_m^v$). Regionalisation is done through the spatial interpolation of the parameters that describe the functions (q, r, u, v). Parameters that depend on soil type, land cover and potential evaporation should be derived locally, as is done in daily models.

Thus, the first innovative step in this dissertation is to describe transition probabilities of a rain-day after a rain-day (p_{01}) and of a rain-day after a dry day (p_{11}) as functions of monthly rainfall. In stochastic rainfall modelling, it is already established knowledge that in many climates occurrence of rainfall depends on occurrence of rainfall on the previous day.

Earlier researchers have used the Markov property to analytically derive statistical properties for the occurrence of rainfall and its persistence. Their work is used to find analytical expressions for the probability density functions of the number of rain-days during a period of a certain length, of frequencies of wet and dry spells and of lengths of wet and dry spells. Until now transition probabilities in stochastic modelling have been described as a function of the season. In this study, however, they are described as a function of the monthly rainfall, and additionally the season if necessary. Using logistic or power functions relating monthly rainfall to the transition probabilities, it becomes possible to express the probability density functions as a function of the monthly rainfall.

In Chapter 4, equations are presented for characteristics of rainfall occurrence that are based on the Markov property. Other researchers have derived these equations to describe the seasonal variability of rainfall occurrence, but here they are transformed into equations that describe the within-month variability of rain occurrence for months for which the total rainfall is available.

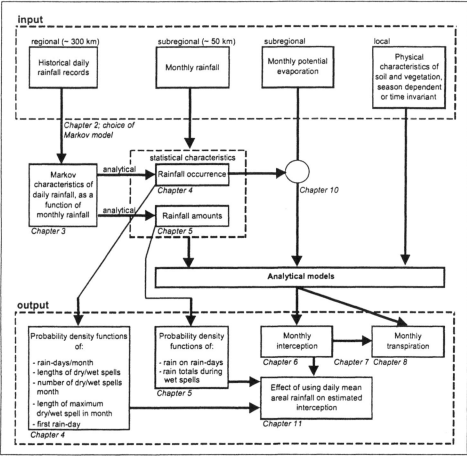

Figure 1.4   Flow chart of approach and dissertation layout.

The second innovative step in this dissertation is to express the scale parameter of the probability of exceedance of rain on rain-days in the transition probabilities of the Markov process. The probability of exceedance of rain on rain-days is modelled as an exponential distribution of which the scale parameter by definition equals the mean rainfall on rain-days (β). The ratio of monthly rainfall to the expected number of rain-days per month, which has been derived through the Markov property, also yields the mean rainfall on a rain-day. In this way it becomes possible to describe both the probabilities of occurrence and persistence of wet spells and the probabilities of a certain depth of rainfall on rain-days as a function of monthly rainfall. This requires only the four parameters that describe the relation between monthly rainfall and the transition probabilities for rain occurrence (q, r, u, v). The spatial variability in these parameters is low, particularly for the two that describe the transition probability of occurrence of a rain-day after a rain-day. Therefore a few daily rainfall series (airports) at a distance of 300 km apart are sufficient to improve monthly relationships considerably.

Chapter 5 starts with the description of the probability density function of the amount of rainfall on a rain-day as an exponential equation. The scale factor of the exponential equation β is the mean rainfall on a rain-day and depends on monthly rainfall, through the Markov process. Data from Harare, Masvingo and Bulawayo are used to test in several ways how well the analytically-derived functions with a monthly time step perform in describing within-month rainfall variability.

The third innovative step in this dissertation is to transform the probability density functions that describe daily rainfall variability into simple equations for monthly moisture fluxes such as interception and transpiration. This is done through analytical transformation of the probability density functions that describe rainfall variability.

In Chapter 6 derivations are presented for monthly interception. Interception, in this dissertation, is defined as all rainfall that evaporates within a day, both from leaves and from the upper soil. As in daily models, a daily threshold is used to represent interception. By combining the knowledge of the number of rain-days in a month and the probability density function of rain on rain-days, an equation for interception is derived.

Chapter 7 addresses transpiration, by disaggregating the month into wet and dry spells of the mean length and the expected rainfall intensity (all derived through the Markov property). In continuous time, transpiration is a linear function of the soil moisture availability, which is the most common approach. A maximum transpiration is reached when the moisture availability is higher than 50% to 80% of the maximum possible, depending on land cover and soil. This spells model performs well, but requires within-month numerical computations with time steps of expected spell lengths.

In Chapter 8 it is shown that the disaggregation of rainfall that was used in Chapter 7 is not necessary. By assuming a constant rainfall intensity during the month, the same accuracy can be reached. A simple model for monthly transpiration is derived that takes note of constrained transpiration during times of less than 50% to 80% of the maximum soil moisture availability. The model has the same input parameters as the daily model and can account for locally-variable physical characteristics of soil and vegetation.

In Chapter 9 the monthly model developed in this dissertation is compared with a number of others, in particular the Thornthwaite-Mather model.

Chapter 10 discusses how potential interception and transpiration in wet and dry spells can be determined from monthly climatic data.

Chapter 11 is a theoretical exploration of the error that is introduced in daily hydrological models which use interpolation between daily rainfall stations as an input. Time series that are derived by spatial interpolation agree to the Markov property, but the rainfall is spread out over more days, and thus interception is comparatively high.

Chapter 12 summarises the methodology and presents the conclusions.

In this dissertation stochastic, statistical and analytical mathematics are employed. Only relatively simple mathematics is used, partly for reasons of analytical tractability. The added value lies in the combination of these approaches, which results in new analytical equations for numerical water resources models with monthly time steps.

2 The Markov Model, a Stochastic Model for Daily Rainfall

2.1 Introduction

Hydrological processes such as interception and recharge are related to daily rainfall, but are not directly proportional to it. When the intention is to model hydrological relations using monthly rain data as input, knowledge about the statistics of daily rainfall distribution is necessary, preferably as a function of monthly rainfall totals. If the statistical characteristics of the few stations for which daily data are available are reasonably similar, these characteristics can be extrapolated to other stations. The statistical characteristics that are looked for should together be sufficient to generate, as a function of monthly rainfall, artificial daily rainfall time series with statistics sufficiently similar to the historical daily time series to reproduce the monthly aggregates of the hydrological processes.

Fluxes such as transpiration and recharge depend on the amount of water in the upper soil. This storage capacity has a 'memory' which is often less than a month. For interception the time scale is smaller: only one day. Therefore, it is important to have knowledge of:
* the number of rain-days in a month,
* the variance in daily rainfall,
* the clustering of rain-days,
* if necessary, the autocorrelation of rainfall amounts.

These factors have been comprehensively studied by many researchers who developed stochastic models for rainfall generation, which will be discussed in the following subsections. In this Chapter the choice of a Markov stochastic rainfall model is explained and an overview is given of previous research regarding Markov stochastic rainfall models.

2.2 Markov models and other stochastic rainfall models

Depending on how the spatial correlation of rainfall is treated, stochastic rainfall models can be grouped into spatial models, multi-site models or single-site models (Waymire & Gupta, 1981 a,b,c). For water resources assessment ideally a spatial model should be constructed. However, the gauging network is usually not sufficiently dense. Recent developments in the theory of Cold Cloud Duration have not progressed far enough yet to describe the daily variability of rainfall (Rugege, 2001). Thus the possibility of using a spatial model drops out. Multi-site models use the cross-correlation structure between the statistical properties of single sites. It is therefore useful to have a look at single-site models first.

When considering temporal single-site rainfall models, two characteristics are important: (i) the rainfall occurrence and (ii) the rainfall intensity. The two are modelled either separately or simultaneously. Usually they are modelled separately, because the high probability of no rainfall distorts statistical distributions.

Box 2-A   Rain-day, dry day or 'trace'?

The definition of a rain-day may be given as a day with measurable rain or as a day with a total rainfall greater than a selected threshold in the order of 0.3 mm/day (Shaw, 1988). Introducing a threshold may be done for several reasons:

- Statistical analyses on occurrence of rainfall may fit better (no references).
- Traces of rainfall are difficult to measure and therefore not always consistently recorded (Stern & Coe, 1984).
- The probability distribution of rainfall amounts minus the threshold value gives a better fit for the probability density function of rain on rain-days (Shaw, 1988).

Stern & Coe (1984) introduce an approach that uses three states in the Markov process, i.e. rain-day, dry day, 'trace'.

In the models for Zimbabwe in this dissertation, a rain-day is defined as any day with recorded rainfall. Previous applications of Markov chain models in the Southern African and Eastern African regions did not define a threshold for a rain-day to be called a rain-day either (South Africa - Zucchini et al., 1992; Tanzania - Stern & Coe, 1984; Kenya - Sharma, 1996a,b). This author tested if a threshold of 0.1 or 0.3 mm/day would make the assumption of a first order Markov chain in Harare more accurate (thus smaller differences between p_{01}, p_{001} etc.). But from graphical analysis the differences appeared insignificant.

In many models, a threshold in the order of 0.1 mm/day is set for days to be considered rain-days. See Box 2-A.

Rainfall occurrence approaches for single sites have been classified into three groups:
- the discrete time series approach,
- the wet-dry spells approach,
- the point process approach.

For the explanation below ample use is made of Hall (1996).

In the discrete time series approach a discrete time step is either wet or dry, although for some models further states are distinguished (see Box 2-A). The probability of a rainfall event can be independent on previous time steps (Bernoulli process) or dependent on them. If it is dependent it is called a discrete autoregressive process. In case this dependency only relates to one day in history, the autoregressive process is called a Markov process, or a Markov chain. Thus, the term 'Markov process' means that the probability of occurrence of an event in a certain time step depends on the state of the system in the previous time step. This is the strict statistical definition, but hydrologists most often use other terminology (see the description of terminology in Box 2-B).

In the wet-dry spells approach, a spell is considered an event. The occurrence of such spells is completely determined by the probability laws of the length of wet and dry spells. This type of model is also known as an alternating renewal model. Roldán & Woolhiser (1982) compared a Markov model and an alternating renewal model and found in favour of the Markov model, on the basis of statistical considerations as well as computational time. Small & Morgan (1986) showed that the transition probabilities of two state Markov processes follow from alternating renewal models that have gamma-distributed dry intervals. (For gamma-distribution, see Appendix B.)

The point process approach employs a more physically-based representation of precipitation and was developed by Kavvas & Delleur (1981) and Waymire & Gupta (1981, a,b,c). The time series are continuous, but each storm is considered to be composed of cells with constant intensities of rainfall. The timing of both the incidence of the storms and the cells within a storm are described by a statistical distribution, generally the Poisson distribution. The depths and durations of the cells in a storm are described by independent statistical distributions. The theory of point processes is applied in Neymann-Scott (arrival of storm cells is randomised) and Bartlett-Lewis rectangular pulse models (interarrival time of storm cells is randomised). For a comparison of the two schemes, see Velghe et al. (1994).

Because of their physical basis and their independence of the level of aggregation, the point process models are most common, or at least most studied by scientists, nowadays. However, the complexity of the point process models makes it difficult to derive analytical relations on an aggregated time scale. Also, most of the routinely collected rainfall data are archived on a daily time scale. If only time series of daily rain sums are available, the calibration of the parameters in point process models is complex. (For the Bartlett-Lewis rectangular pulses model a method was developed by Bo, Islam & Eltahir, 1994).

However, this author questions whether, in climates where convection is the dominant rainfall-producing mechanism, point process models are indeed more physically-based than Markov models. As mentioned, the different point process models (Poisson White Noise, Rectangular Pulse Poisson, Neymann Scott Rectangular Pulse, Bartlett-Lewis) make use of Poisson arrival of storm origins. Point process models have been developed in climates where convection is not dominant (May and June in Denver, Colorado, U.S.: Rodriguez-Iturbe et al., 1987; Velghe et al., 1994; Blackpool in the United Kingdom: Cowperthwait, 1991). When convection is the dominant rainfall-producing mechanism, as is the case in Zimbabwe, rainfall usually starts late in the afternoon, when the soil warmed sufficiently. Poisson arrival of storm origins does not take account of this. It can be expected that a Markov model with, in the event of rain, a random arrival time that is distributed around, say, 4 o'clock in the afternoon, is physically more accurate. The current author does not address this hypothesis further and it is not used in the choice of a Markov model.

Markov models have been proven applicable in many semi-arid and also in some more humid climates, see Table 2-1. Markov models on rainfall occurrence have the advantage that they are based on the common observation time step of one day. Their disadvantage is that the time of occurrence during the day can only be modelled using complex disaggregation techniques (Hershenhorn & Woolhiser, 1995). For semi-arid areas where convection is the most important rainfall-producing mechanism, neither advantage nor disadvantage are serious, because rainfall from convection usually occurs at the end of the afternoon. On a monthly time scale, which is the objective of this study, the exact occurrence of the storm in the day has little relevance, although it may influence interception.

Box 2-B   Markov processes: explanation of terminology

In a two **state** Markov process, the process is in either one of two states at a certain time step. For Markov rainfall models, these two states are generally 'dry day' and 'rain-day'. The threshold between dry day and rain-day may be higher than 0 mm/day, see Box 2-A. In a rainfall model with a three state process distinction is made between 'dry day', 'day with trace rainfall' and 'rain-day'. An x-state Markov model classifies days into an x number of ranges of daily rainfall amounts.

A Markov process only depends on the state of the process in the preceding time step. This is the correct mathematical definition (e.g. definition by Hoel, Port & Stone, 1972), which is used in this dissertation. However, many hydrologists refer to autoregressive processes of orders higher than one as Markov processes. In such a case they refer to a Markov process as a first-order Markov process. If there is reference to a 'second-order Markov process', an autoregressive process is meant that depends on the two preceding time steps. And so on. For reasons of readability, in the following chapters **a two state Markov process is simply referred to as a Markov process**.

Most Markov rainfall models in the literature are **nonhomogeneous Markov processes**, which means that the conditional probabilities vary with time. For rainfall models the probabilities usually depend on the time in the season. This can be a harmonical dependency (e.g. Clarke, 1998) or an empirical derived dependency (e.g. Zucchini, Adamson & McNeill, 1992).

Bárdossy & Plate (1991) make use of a **semi-Markov process**. The Markov processes are used for transition between states. However, the duration of states is not of a fixed length, but is drawn from a probability density function. In Bárdossy & Plate's work the states do not directly refer to amounts of rainfall, but to the occurrence of certain atmospheric circulation patterns. Each circulation pattern has a certain probability of occurrence of rain, with a weak persistence as long as the circulation pattern stays the same. Wilby, Greenfield & Glenny (1994) used a proper Markov model for simulation of the seven Lamb's Weather Types. Each Lamb Weather Type has its own probability of rainfall. Both approaches follow the line of thought of the **Markov renewal model** of Foufoula-Georgiou & Lettenmaier (1987), who use the Markov process for states with different probability of rainfall, but who do not attach a physical meaning to these states.

Smith (1987) introduces for rainfall modelling a **Markov-Bernoulli process**. A Bernoulli process just means that at a certain time step there is a certain probability of success (rain), without dependence on previous time steps. In the Markov-Bernoulli process, the success probability is a seasonally varying Markov chain. Thus, the probability of rain does not depend on the occurrence of rain on the previous day, but on the probability of rain on the previous day. In this simple way complex stochastic models can be constructed. Smith shows that in this fashion a discrete analog of the Neymann Scott cluster can be derived.

Markov rainfall models do not only exist at time steps of days, but also of hours (e.g. Hutchinson, 1990), months and years (Gregory et al., 1993). In hydrology, not only rainfall models use Markov processes. Many stochastic runoff models also follow Markov processes. For example, Lu & Berliner (1999) developed a Markov runoff model that switches between states of rising, falling and normal. In September 2000, in the database Water Resources Abstracts, 670 references related to 'Markov', of which 160 were for rainfall models. However, as mentioned above, many of the models that are referred to as Markov models, are autoregressive models of a higher order than 1.

Comparisons of the performance of Markov models and point process models are rare. The only comparison that this author is aware of favours the Markov model: Madamombe (1994) found that for the semi-arid environment of Tanzania, the Markov process performed better on a daily time scale than the Neyman Scott Rectangular Pulses model. Madamombe compared two models from two model classes and his conclusion is partly based on their performance in the dry season, which is not as relevant for water resources assessment. However, the conclusion seems quite logical for a climate where most rainfall is the result of convection, as is the case in most semi-arid countries. In such a climate a wet spell may persist for several days, but it consists of separate storms.

Clarke (1998) identified three methods to derive practical results on the seasonal variability of certain probabilities from stochastic rainfall occurrence models:
- Analytical, in which a formula is derived to give the required result in terms of the model parameters. In practice this is very difficult for all but the simplest of models;
- Recurrence relations based upon autoregressive processes;
- Computer simulation.

Thus, the advantage of the discrete autoregressive processes is that recurrence relations can be calculated. It is explained in Chapter 3 what this implies. Several authors have used the recurrence relations of autoregressive processes, in particular of the Markov process, to describe seasonal variabilities in rainfall (e.g. Gabriel & Neumann, 1962; Stern & Coe, 1982; Sharma, 1996a).

For the above reasons, the properties of stochastic daily rainfall models based on autoregressive processes seem most suitable in the pursuit of the objective of this dissertation, namely to improve monthly water resources models by including the statistical characteristics of daily rainfall.

2.3   Markov-based daily rainfall models

Gabriel & Neumann (1962) apparently were the first to use a Markov process for the occurrence of rain-days. Their model was based on mid-winter data from Tel Aviv. They determined conditional probabilities independent of annual or monthly rainfall and constant for the mid-winter period.

Table 2-1 shows that stochastic daily rainfall models based on autoregressive processes have been developed for locations worldwide. The table is illustrative, but by no means complete. In most cases the models have seasonally-varying conditional probabilities, often modelled with harmonics.

In South Africa, in the same region as Zimbabwe, Zucchini et al. (1992) produced Markov models for occurrence of daily rainfall, calibrated for 2,550 stations. They used two harmonics to describe the variation of conditional probabilities through the season.

If the probability of rainfall occurrence depends on the amount of rainfall on the previous day, a model with more than two states is necessary. Haan et al. (1976) described such models. Other authors (e.g. Gregory et al., 1993) confirmed this. Of course, the fit is bound to improve, because more parameters are available. However, as a consequence the calibration process becomes more complicated and more data are required. Moreover, the analytical relations derived in this dissertation would be far less transparent. Therefore, a two state process is preferred here.

The introduction of a threshold of say 0.3 mm/day, may result in better fits without any consequence for the derivation of the analytical relations. But, if thresholds are too high, the persistence disappears and zeroth-order models are sufficient (Smith, 1987). In the next Chapter it will be shown that a two state Markov process without a threshold is accurate for Zimbabwe.

It is interesting to note that Foufoula-Georgiou & Georgiou (1987) show that an x-state autoregressive process of order k can be regarded as a first-order chain with x^k states. Gregory et al. (1992) investigated the opposite: A first-order many states Markov process can also be described by a two state higher-order autoregressive process. The required order depends entirely on the values in the transition matrices (see Appendix A).

The central thought that shaped this dissertation is that probabilities of occurrence of a rain-day after respectively a dry or a rain-day are larger in months with more rainfall. Wilks (1989) used the same thought. He fitted Markov models of daily time step to dry, near-normal and wet subsets of monthly total precipitation. The purpose was to use forecasts of 30-day precipitation totals, as routinely issued by the United States Climate Analysis Center, as an input to Monte Carlo simulation at a daily time step. Instead of using the forecasted 30-day precipitation total, Wilks only made the distinction between the three subclasses for each month in the year.

To the author's knowledge, only Mulligan & Reaney (1999) used Markov chains for disaggregation from monthly to daily data. However, they used a five state seasonally-dependent Markov process and, starting with the last day of the previously generated month, continued to generate monthly series of daily rainfall until a monthly total was obtained that was close to the monthly value to be disaggregated. For reproduction of daily rainfall amounts a Monte Carlo approach is used, after Bárdossy (1998). Obviously, with the set conditional probabilities there is no unique solution for the disaggregation of the month. The method does not show at all how probable it is that such a daily time series occurs. Therefore this method is not suitable for deriving analytical equations of interception and transpiration at a monthly time step.

To take into account that different months in the year with the same amount of rainfall can show different daily rainfall patterns, it may be necessary to distinguish between different months in the Markov analysis. Most stochastic rainfall models have such seasonally-varying conditional probabilities, often based on harmonics (see Table 2-1, e.g. Zucchini et al., 1992; Smith & Schreiber, 1974; Clarke, 1998; Wilks, 1989). But to derive conditional probabilities as a function of monthly rainfall for each of the six months in the rainy season separately, would require about six times more data in order to obtain the same accuracy in the estimates. As average monthly rainfall totals vary over the rainy season, the variation in conditional probabilities over the rainy season is already in some way represented in the conditional probabilities as a function of monthly rainfall. In Chapter 3 it will be shown that in the case of Zimbabwe it is not necessary to make a distinction between months in the year. This does however not apply to all locations worldwide.

Stochastic daily rainfall models based on the Markov chain do not preserve the variability in monthly rainfall very well (Buishand, 1977; Woolhiser et al., 1993, who refer to Zucchini & Adamson, 1984). Mimikou (1984) found that for Greece the monthly total of rain-days was best described through an autoregressive stochastic model. The Markov transition probabilities were adjusted until an equal number of rain-days was obtained. Thus, Mimikou uses the total number of rain-days within a month, which in her case is stochastically generated, and obtains a non-unique set of transition probabilities, in a similar way to Mulligan & Reaney (1999). The problem of Markov models not conserving the variability of monthly rainfall is not relevant to the method in this dissertation, because historical monthly rainfall amounts are used as input. Yet the fact that the variability of monthly rainfall is not preserved, shows that care should be taken in the use of stochastically-generated rainfall series as input in water resources models.

For the amounts of rainfall on rain-days, gamma, Weibull or exponential functions are used in the literature (see Table 2-1). The exponential distribution is most appropriate for the purpose of analytical derivations, but does not represent the relatively high probability of low rainfall amounts. In Chapter 5 the choice of an exponential distribution is argued.

2.4   Conclusions

The use of the properties of two state Markov models for daily rainfall occurrence is most suitable to upgrade equations in water resources models at a monthly time step. The arguments are:

- Markov models have suitable mathematical properties to derive probability density functions directly from the model parameters without simulation.

- To calibrate Markov models, daily rainfall records are sufficient. Event-based models would need rainfall records in continuous time or at least the application of complex methods of disaggregation.

- Two state Markov models have been applied at many locations worldwide.

- Two state Markov models are simple and therefore analytical derivations are transparent.

Table 2-1   Stochastic models for daily rainfall, based on discrete autoregressive processes.

region	location	occurrence of rainfall using autoregressive processes			rain amount on rain-days ● Weibull ■ gamma ▲ exponential or mixed exponential	reference
		order (1st = Markov chain)	states	thres-hold (mm /day)	seasonal variability of conditional probabilities		

region	location	order (1st = Markov chain)	states	thres-hold (mm /day)	seasonal variability of conditional probabilities	rain amount on rain-days	reference
AFRICA	Southern Africa, 2,550 meteorological stations	1st	2	0	2 harmonics fitted to conditional probabilities on a daily basis.	● Weibull, seasonally varying mean, constant C_v. No autocorrelation.	Zucchini, Adamson & McNeill (1992)
	Tanzania (Morogoro)	1st	2	0	4 harmonics fitted to conditional probabilities on a 5-day basis.	■ gamma, with $\kappa = 0.768$ or κ bias reduced = 0.521, dependent on rainfall occurrence on previous day.	Stern & Coe (1984)
	Tanzania (Morogoro and Dodoma. The latter is more arid)	1st	2	0	Conditional probabilities calibrated for each month in rainy season.	Not applicable. Only rainfall occurrence was studied.	Madamombe (1994) The performance of the Markov model is better than a Neyman-Scott rectangular pulse model
	Kenya, semi-arid (Kibwezi) and semi-humid (Kabete)	1st	2	0	Conditional probabilities are constant, but different for long and short rainy season. The persistence is greater for semi-humid station.	● Weibull, for total rain of wet spell.	Sharma (1996a)
	Nigeria (Kano $\overline{P}$ =850 mm/y mm/y, Samaru, Bida, Enugu, Calabar $\overline{P}$ =3000 mm/y)	1st: 2nd equal perfor-mance	2	0.3/2/5	Conditional probabilities calibrated for each month in rainy season.	Not applicable. Only rainfall occurrence was studied.	Jimoh & Webster (1996)
	Zimbabwe	1st	2	0	Dependent on monthly rainfall, not on seasonal variability	▲ exponential, dependent on monthly rainfall, not on seasonal variability.	De Groen, this dissertation

region	location	occurrence of rainfall using autoregressive processes				rain amount on rain-days ● Weibull ■ gamma ▲ exponential or mixed exponential	reference
		order (1st = Markov chain)	states	threshold (mm/day)	seasonal variability of conditional probabilities		
ASIA	India (Hyderabad)	2nd	2	0.1	See Tanzania, same authors.	■ gamma	Stern & Coe (1982)
	India (Nagpur)	5th	2	0	See Tanzania, same authors.	■ gamma.	Stern & Coe (1984)
	Vietnam (Red River Delta)	1st	2	1	Conditional probabilities calibrated for each month in rainy season.	lognormal.	Binh, Murty & Hoan (1994)
MIDDLE EAST	Jordan (Irbid)	1st	2/3	'trace' is 3rd state.	See Tanzania, same authors.	■ gamma, seasonally varying mean rainfall, constant κ, dependent on rainfall occurrence on the previous day.	Stern & Coe (1984)
	Israel (Tel Aviv)	1st	2	0.1	Historical monthly average, but constant in midwinter (Dec-Jan-Feb).	Not applicable. Only rainfall occurrence was studied.	Gabriel & Neumann (1962), Katz (1981, referred to by Jimoh & Webster, 1996)
		2nd	2	0.1	Ibid.	Ibid.	Gates & Tong (1976) To determine the order, the Akaike Information Criterion was used.
EUROPE	United Kingdom (Manchester and Liverpool)	1st	2	0.1	Nov-Feb constant and March-June constant.	Ibid.	Ibid.
AUSTRALIA	Central Australia 6 stations (125 mm/y – 375 mm/y)	1st	2	2.5 mm/pentad	0 or 1 harmonic.	Ibid.	Fitzpatrick & Krishnan (1967). N.B. applied on pentads, not on days.

region	location	occurrence of rainfall using autoregressive processes				rain amount on rain-days ● Weibull ■ gamma ▲ exponential or mixed exponential	reference
		order (1st = Markov chain)	states	threshold (mm/day)	seasonal variability of conditional probabilities		
NORTH AMERICA (all USA)	Indiana (Indianapolis), Missouri (Kansas), Wyoming (Sheridan), Florida (Tallahassee)	1st	2	0	2 harmonics, determined by Maximum Likelihood method.	▲ mixed exponential (= sum of two exponentials), seasonally-varying parameters, no correlation with previous days.	Woolhiser & Pegram (1979)
	Arizona (12 stations), California (9), Nevada (1), New Mexico (5) Texas (Austin)	1st	2	dependent on station	2 harmonics, p_{01} depending on Southern Oscillation index (SOI).	▲ mixed exponential, parameters depending on Southern Oscillation Index.	Woolhiser, Keefer & Redmond (1993)
	Arizona (Sonoran and Chihuahuan region)	1st	2/3	0	Conditional probabilities calibrated for each 10-day period in the season.	▲ exponential, scale parameters for each 10-day period in the season.	Todorovic & Woolhiser (1975)
		1st	2	0	For each day in the season, conditional probabilities were derived at three stations (55-73 years of data). The daily values are smoothed through a 5-day moving average.	Not applicable. Only rainfall occurrence was studied.	Smith & Schreiber (1974)
	>100 stations in USA	1st/2nd /(3rd) dependent on season	2	0.25	1st order in summer, 2nd order in winter, on basis of Akaike index. This was later questioned by Katz (1981) as Bayesian Information Criteria gave 1st order only.	Ibid.	Chin (1977)
SOUTH AMERICA	Amazon, Brasil	2nd	2	0.1	2 harmonics to represent variation through year.	■ gamma, $\kappa = 0.771$	Clarke (1998)

3 Occurrence of Rain-Days as a Markov Process

3.1 Introduction

In this chapter it is shown that for Zimbabwe the probability of occurrence of a rain-day is conditional on the occurrence of rain on the preceding day. For different class intervals of monthly rainfall, the probabilities of this Markov property and the probabilities of exceedance of rainfall on rain-days can be derived.

3.2 Notations

As explained in the previous chapters, if a time series has the Markov property it means that the probability of occurrence of a certain state depends only on the state in which the system was during the previous time step. It was decided that a two state Markov process at daily time steps would be used, with a dry day ($P_t = 0$) and a rain-day ($P_t > 0$) as the two states.

In this dissertation a rain-day is defined as a day with recorded rain, not as one with rain above a certain threshold (see Box 2-A). The subscripts 1 and 0 are used for respectively a rain-day or a dry day. Consequently, the transition probability of a rain-day following a rain-day is represented by p_{11} and the transition probability of a rain-day after a dry day is represented by p_{01} (annotations similar to Grimmett & Stirzaker, 2001).

Thus, the transition probability p_{11} of a two state Markov process agrees to:

$$p_{11} = \mathbf{P}\langle P_t > 0 | P_{t-1} > 0 \rangle$$

$$
\begin{aligned}
&= \mathbf{P}\langle P_t > 0 | P_{t-1} > 0, P_{t-2} > 0 \rangle \\
&= \mathbf{P}\langle P_t > 0 | P_{t-1} > 0, P_{t-2} = 0 \rangle \\
&= \mathbf{P}\langle P_t > 0 | P_{t-1} > 0, P_{t-2} = 0, P_{t-3} = 0 \rangle
\end{aligned}
$$

rain-day after rain-day   Eq. 3.1

.....

The probability of a rain-day after a dry day is similar:

$$p_{01} = \mathbf{P}\langle P_t > 0 | P_{t-1} = 0 \rangle$$

$$
\begin{aligned}
&= \mathbf{P}\langle P_t > 0 | P_{t-1} = 0, P_{t-2} > 0 \rangle \\
&= \mathbf{P}\langle P_t > 0 | P_{t-1} = 0, P_{t-2} = 0 \rangle \\
&= \mathbf{P}\langle P_t > 0 | P_{t-1} = 0, P_{t-2} = 0, P_{t-3} = 0 \rangle
\end{aligned}
$$

rain-day after dry day   Eq. 3.2

.....

Box 3-A   The climate of Zimbabwe

The rainfall in Zimbabwe is strongly related to the seasonal fluctuations of the Inter-Tropical Convergence Zone (ITCZ), the zone where the airstreams originating in two hemispheres meet. The ITCZ is not one permanent, globe-encircling region, where airstreams are always convergent, but a complex, ever changing band of growing and disintegrating convergences (description by Buckle, 1996). The position, width and depth varies geographically, seasonally and even daily. The ITCZ moves with the sun, southwards at the beginning of the rainy summer season and northwards at the end of it. As a consequence, the rainy season in the north starts earlier and finishes later than that in the south.

The convergence within the ITCZ induces convection, which is the movement of air upwards due to warming at ground level. Convection accounts for perhaps 90% of the Zimbabwean rainfall, although not all of this is related to the ITCZ (Torrance, 1981). Convection can result in the generation of huge cumulonimbus clouds, from which energy and rain are released during thunderstorms. The upward movement of air is compensated for by the downward movement of air (subsidence) in areas around it. Convection thus only occupies a maximum of 10% of the area, with dimensions in the order of 5 to 10 $(km)^2$. Therefore, the spatial scale of daily rainfall is very small. Interception changes the energy balance locally and so induces convection, causing persistence in the occurrence of rain-days (Taylor, 1997).

Apart from the ITCZ, Indian Ocean cyclones influence Zimbabwean rainfall, most violently illustrated by the cyclone Eline that struck Mozambique, southern Zimbabwe and northern South Africa at the beginning of 2000. Indian Ocean cyclones are frequent phenomena, but they normally do not penetrate so far inland. The cyclone season is usually from December to April (Makarau, 1995). Most storms curve back southward in the Mozambique channel and dissipate at higher latitudes (Buckle, 1996). Their influence on the weather in Zimbabwe varies, depending on the month and their strength and precise track. Increased rain is usual within a 100 km radius of the cyclone, but further afield the effects are sometimes completely the reverse. Buckle (1996) and Matarira & Jury (1992) mention the example of cyclones that curve south near Madagascar and cause an influx of dry, upper air into Zimbabwe and Mozambique, resulting in persisting dry spells. Torrance (1981) mentions an example where the remains of a cyclone cause a deep low-pressure area, which brings strong winds that enhance orographic rainfall.

On the other hand, migratory anticyclones (diverging air) along the southeast coast play a role. The occurrence of long wet spells at many locations in Zimbabwe is often related to a lower northerly and an upper easterly air flow, which corresponds with a tropical low over Zambia and an anticyclone off the Indian Ocean coast (Matarira & Jury, 1992).

Annual rainfall and the start and end of the rainy season are subject to 'teleconnections' (Matarira & Flocas, 1989). As a consequence, annual rainfall shows a periodicity of 13-18 years (Tyson, 1986). Torrance (1981) has suggested using the height of the Nile flood in August and September to forecast the rainy season. Many recent research efforts concentrate on the forecasting values of El Niño and the Southern Oscillation (ENSO index) (Matarira, 1990; Unganai, 1996), while the Nile flow is also related to El Niño (Eltahir, 1996). Although significant correlation exists, the predictive value of forecasts is limited. While the uncertainty of the forecasts is still considerable, however, the public has been informed of the most suitable planting dates for their crops on the basis of such forecasts. In this way, a well intended but uncertain forecast has made the economic system less robust than when it relied on the known annual variability in rainfall. In this dissertation, historical monthly rainfall data are used as input and the annual variability is thus represented.

In the same way, the probability of two successive dry days is p_{00} and of a dry day after a rain-day p_{10}. Because a day is either a dry day or a rain-day,

$$p_{01} + p_{00} = 1 \qquad \text{Eq. 3.3}$$

and

$$p_{11} + p_{10} = 1 \qquad \text{Eq. 3.4}$$

Therefore, all relationships can be expressed as functions of two transition probabilities, p_{11} and p_{01}.

To simplify notations, a parameter C is introduced

$$C = p_{11} - p_{01} \qquad \text{Eq. 3.5}$$

3.3   Harare, Masvingo and Bulawayo

The method presented in this dissertation uses daily records at a few locations to derive statistical parameters that describe the variability of rainfall within the month. For these locations the monthly models have been compared with aggregated daily models, to evaluate whether they accurately represent the hydrological processes. The stations used are Harare, Masvingo and Bulawayo,which have been chosen because in 1998 their data were readily available in digital format. The daily data which have been digitised since then are only used to verify the modelling approach (Section 3.7). This Section briefly describes the differences in climate between Harare, Masvingo and Bulawayo. In Box 3-A a general description of the climate of Zimbabwe is given. Table 3-1 presents the details of the stations and Figure 1.1 shows the locations on the map.

The mean annual rainfall at Harare (790 mm/y) is on average higher than that at Masvingo (640 mm/y) and Bulawayo (630 mm/y). Harare is further north than Masvingo and Bulawayo. In general, in Zimbabwe average annual rainfall increases from south to north. The northern parts are more often in the influence area of the Inter-Tropical Convergence Zone (ITCZ, see Box 3-A).

Additionally, winds of a northerly origin (Congo air from the northwest, monsoon air from the northeast) are much more moist than winds from the south (southeast trades) (Torrance, 1981). As a consequence of the ITCZ-related activities, rain occurs more often at Harare than at Bulawayo and Masvingo, despite the same amount of moisture in the air (Torrance, 1975; De Groen & Savenije, 1996). Additionally, conditions in Bulawayo and, to a lesser extent, Masvingo are far more influenced by the occurrence of a high pressure zone in the upper air above Botswana. This Botswanean Upper High causes an anticyclonic movement of air masses in the region (air diverges anticlockwise) and therefore has a rain-inhibiting stability (Torrance, 1975; Makarau, 1995).

Of the three stations, Harare is at the highest altitude, followed by Bulawayo. Gradients in rainfall amounts due to the movement of the ITCZ or any other north-south influence are distorted by the strong connection between altitude and rainfall. There is much more rainfall at high altitudes than lower down, both on small and on large spatial scales (Torrance, 1981; Abebe, 1996; Laing, 1973). Masvingo and Bulawayo are at lower altitudes than Harare and have on average less rainfall. Masvingo is in an area that is influenced by guti, drizzle caused by orographic processes prevailing on southeasterly slopes (Torrance, 1981).

Table 3-1   Details of the stations from which data are used throughout the dissertation.

	Harare airport	Masvingo	Bulawayo Goetz
mean annual rainfall (mm/y) in years from which data are used	790	640	630
latitude	17°55'S	20°04'S	20°09'S
longitude	31°06'	30°52'	28°37'
altitude (m)	1,470	1,100	1,340
mean number of thunderstorm days (Torrance, 1981)[1]	73	54	63
data period	1959/1960 - 1992/1993	1951/1952 - 1996/1997	Jan 1930 –1994/1995
missing data	1982/1983, 1989/1990	Feb 1959 - Nov 1965, 1978/1979, 1987/1988, 1988/1989	Feb 1932, Apr 1939, Apr 1942, Apr 1947, Jan 1956, Apr 1987, Mar 1990, Apr 1991

[1] By international agreement a thunderstorm day is defined as a local calendar day on which thunder is heard, irrespective of the occurrence of rainfall at the location of measurement (Kreft, 1972).

3.4   Validity of Markov property for different rainfall classes

Figure 3.1 gives charts of different monthly rainfall classes for Harare data. In each chart the probability of a rain-day is plotted against the number of days in the preceding dry or wet spell that are taken into account. Thus, the upper lines in each graph give estimates for respectively p_{11}, p_{111}, p_{1111}, which are well above the lower lines with estimates of p_{01}, p_{001}, resp. p_{0001} etc.. Figures 3.2 and 3.3 show the same for Masvingo and Bulawayo respectively. If rainfall occurrence would agree to a perfect Markov process, the lines would be horizontal. Figures 3.1, 3.2 and 3.3 also show that for higher monthly rainfall amounts, the transition probabilities of the occurrence of a rain-day increase. This is treated in the next Section. Here it is first discussed whether daily rainfall occurrence for the three stations can be considered a Markov process.

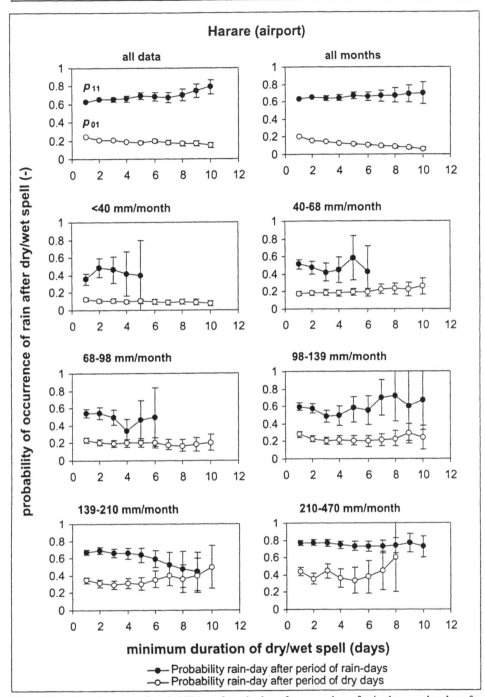

Figure 3.1   Development of probabilities of a rain-day after a number of rain-days or dry days for Harare for different monthly rainfall classes. In each chart the line of higher probability (•) represents the probability of rain after a wet spell (respectively p_{11}, p_{111}, p_{1111} etc.). Similarly the line of lower probability (o) gives the probability of rain after a dry spell (resp. p_{01}, p_{001}, p_{0001} etc.). The graph 'all data' refers to series of data where, in contrast to 'all months', occurrence of rain on the first days of the month, which is conditional on the last days of the previous month, is also taken into account. The bars refer to the 95% confidence limits of the estimated transition probability (see Box 3-B).

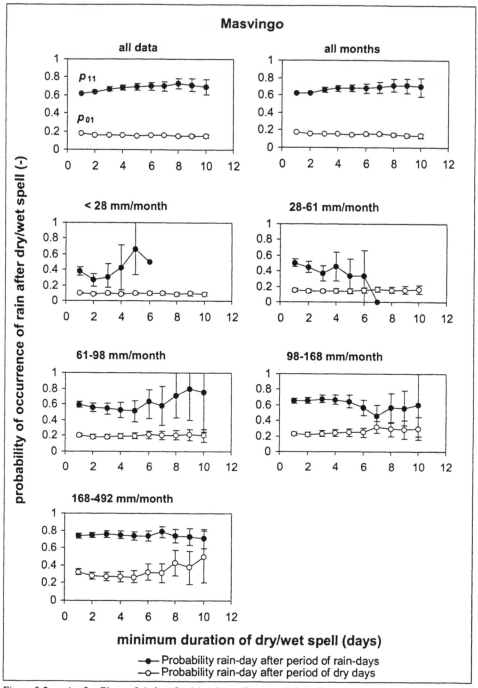

Figure 3.2   As for Figure 3.1, but for Masvingo. Fewer rainfall classes are shown, because fewer historical data are available.

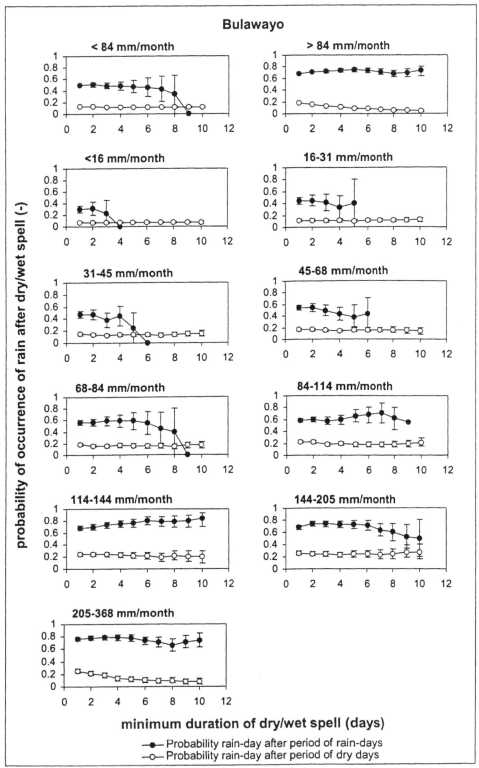

Figure 3.3   As for Figure 3.1, but for Bulawayo. More rainfall classes are shown, because more historical data are available.

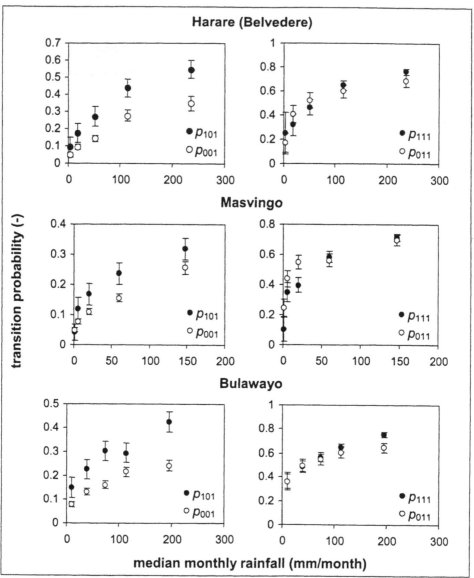

Figure 3.4   The median of a monthly rainfall class plotted against the probability of a rain-day after a dry day preceded respectively by a dry day p_{001} or a rain-day p_{101} (graphs left) or against the probability of a rain-day after a rain-day preceded by a rain-day p_{111} or a dry day p_{011} (graphs right), with the 95% confidence limits.

Because Figures 3.1, 3.2 and 3.3 are more or less horizontal, a two state Markov property seems appropriate.[6] However, the left-hand graphs in Figure 3.4 show that for the probability of rain-days after dry days the state of two days earlier matters. In other words, p_{101} and p_{001} differ significantly for each rainfall class at all locations. The right-hand graphs in Figure 3.4 show that for the probability of rain-days after rain-days, the state of two days earlier is not important.

---

[6] Strictly speaking, horizontal lines do not prove that a process is perfectly Markovian. In such a case other histories also need to be compared. E.g., the occurrence of rainfall may be related to the previous day and to three days before, but not to two days before ($p_{001} = p_{101}$, but $p_{0101} \neq p_{1101}$).

However, a second-order autoregressive process instead of a Markov process increases the number of parameters from two (p_{01}, p_{11}) to four (p_{001}, p_{101}, p_{011}, p_{111}). This decreases the degrees of freedom, with the risk of overfitting. A commonly used fit measure that rewards good fit but penalises loss of degrees of freedom is the Akaike Information Criterion (*AIC*):[7]

$$AIC = n \ln(\hat{\sigma}^2) + 2 * \omega \qquad\qquad \text{Eq. 3.6}$$

where

n is the size of the sample, in this case the number of days,

$\hat{\sigma}^2$ the estimated variance of the residual series,

ω the number of parameters, thus $\omega = 2$ for the Markov process and $\omega = 4$ for the two state autoregressive process.

Chin (1977) used the *AIC* to determine the optimum model order for the more than one hundred stations that he modelled in the USA (see Section 2.3). An alternative to the *AIC* is the Bayesian Information Criterion (Schwarz, 1978, referred to by Jimoh & Webster, 1996). This Criterion is more punitive to loss of degrees of freedom. Therefore if the *AIC* prefers a Markov process over a second order autoregression, the Bayesian Information Criterion will certainly prefer the Markov process.

For Harare (Belvedere) the *AIC* was determined for all months separately, for transition probabilities that were fitted to the individual months and to the classes of months.[8] On average the Markov process had a lower *AIC* than the second-order autoregressive process. However, the differences are rather small.[9] The *AIC* is also determined for classes as a whole.[10] In this case, the differences in *AIC* between Markov and second-order autoregressive processes are negligible. This conclusion differs from Jones & Thornton (1997) who considered the second order and third order for Harare non-negligible.[11]

Jimoh & Webster (1996), using data from Nigeria, showed that frequency duration curves for wet spells are more suitable to determine the most appropriate order than the *AIC* and the Bayesian Information Criterion. Frequency duration curves for wet and dry spells will be shown in the next chapter and they confirm that a first order Markov chain is appropriate. Yet in the frequency duration curves it is more difficult to recognise the need for a higher order than it is in Figures 3.1, 3.2 and 3.3. In this dissertation the judgement is not only based on wet spells, as suggested by Jimoh & Webster, but also on dry spells, which are equally important.

---

[7] Here the definition of the *AIC* as by Stewart (1991) was used. The *AIC* was developed by Akaike (1972).

[8] If a rain-day occurred after a dry day, the residual is taken as 1-p_{01}. For a dry day after a dry day the residual is 0-p_{01}. Similarly, residuals of rain-days after rain-days are 1-p_{11} and dry days after rain-days 0-p_{11}. For the second order autoregressive process, the history of the previous two days was taken into account and residuals are related to p_{111}, p_{011}, p_{001} or p_{101}. If the *AIC* is determined for each month separately, $n = n_m$ is the number of days in the month.

[9] For the fits to the individual months, the average difference (-0.03) is almost four times smaller than the standard deviation of the differences (0.12), thus the difference is insignificant. For the fits to the classes (i.e. averaged fits of the individual months are used), the *AIC* was significantly in favour of the Markov process, with an average difference (-0.12) three times as large as the standard deviation (0.4).

[10] If the *AIC* is determined for a class interval as a whole, n is the number of days in the class.

[11] Jones & Thornton derive for Harare that the differences p_{111} - p_{011} = p_{101}-p_{001}= 0.28 throughout the year and that the differences p_{1111}- p_{0111} = p_{1011}- p_{0011}= p_{1001}- p_{0001} = p_{1101}- p_{0001} = 0.13.

Sharma (1996a) suggests using the cumulative density function of the longest dry and wet spells to evaluate the performance, because his objective is to design a rainfall harvesting system.

As explained in this Section, when the predictive value of a stochastic model is the criterion, the differences in performance between Markov processes and second-order autoregressive processes are small and mostly in favour of Markov processes. Therefore, the choice of a Markov model appears justified. This justification is a separate issue from the preference for a Markov process in order to keep the analytical expressions that are the objective of this dissertation as simple as possible.

3.5   Relationship between monthly rainfall and transition probabilities

In the previous Section it was shown that the Markov property was valid for different classes of monthly rainfall amounts. But, as one would expect, Markov properties vary with monthly rainfall amounts. In a month with more rainfall the probability that a dry or a rain-day is followed by a rain-day is greater, resulting in higher values of p_{01} and p_{11}.

The next step is to express the transition probabilities as a function of the monthly rainfall. The individual monthly estimates of the transition probabilities as a function of the monthly rainfall are very scattered, which is illustrated by the dots in Figures 3.5 and 3.6. However, if the months are clustered in classes of monthly rainfall that are each considered as one sample, there is a clear relation for all locations between the monthly rainfall and the transition probabilities (see the squares in Figures 3.5 and 3.6). Instead of using the estimated probabilities for individual months,[12] the class means were used to fit the relationships. The disadvantage is that the information of the individual months is clustered around the median monthly rainfall of a class. However, the large advantage is that extreme cases, which are very uncertain, are weighed less heavily. For example, in a month with only one dry day, the estimate of p_{01} will be 1, but this is based on only one transition from a dry day to a following day. Therefore, the months with relatively many dry days in comparison with the monthly rainfall should weigh heavier in the determination of a relationship for p_{01}. Similarly, the months with relatively many rain-days should weigh heavier for p_{11}. This is implicit when using class means.

At least five classes were considered necessary to determine the relation with monthly rainfall. Additional classes would yield additional points for the regression of relations, but on the other hand each point would be less accurate. Classes of a sample size of about forty months were chosen, which produced for Masvingo, with

---

[12] The histograms of the transition probabilities for different monthly rainfall classes show non-symmetric distributions, in particular for very low and very high monthly rainfall. With low monthly rainfall, there are many occasions in the analysis when $p_{11} \approx 0$, as the few days on which rain occurred are not followed by a rain-day. With very high monthly rainfall, the probability that a dry day is followed by a rain-day is very high. It follows from this non-normality that minimisation of mean square errors is not appropriate. The different histograms indicate that the likelihood functions are different for different amounts of monthly rainfall. This makes application of the maximum likelihood function very cumbersome. To fit the relationships when using the individual months, minimisation of absolute errors appears the most appropriate way.

the shortest historical record, five classes, for Harare six and for Bulawayo nine. As the variance in monthly rainfall is greater for quantiles of months with higher rainfall, the class intervals increase with monthly rainfall. For the class of lowest rainfall a larger sample size was chosen (fifty months), but this class still had the smallest class interval width. The median monthly rainfall of each class sample is chosen as the representative amount. The mean monthly rainfall in a class and the centre of the class interval lie generally on either side of the median of a rainfall class.

For the relationship between monthly rainfall and p_{01}, there is a constraint to the regression, because a month with 0 mm/month rain has no rain-day, so $p_{01}(P_m = 0) = 0$.

A power function appears to fit the relation between monthly rainfall and rainfall transition probabilities. The transition probability p_{trans} (either p_{01} or p_{11}) is thus modelled as:

$$p_{trans} = \theta_1 (P_m)^{\theta_2}$$ Eq. 3.7

where P_m is the monthly rainfall (mm/month) and θ_1 and θ_2 are factors derived through calibration. (To get a balance in units θ_1 would need to be in $(month/mm)^{\theta_2}$.)

The disadvantage of a power function is that, for very extreme monthly rainfall, the modelled probability can become higher than 1, which is impossible. It is theoretically more correct to use a function of the logistic form[13]

$$p_{trans} = \frac{1}{1 + \theta_1 (P_m)^{\theta_2}}$$ Eq. 3.8

(personal communication from Prof. A. Bárdossy, University of Stuttgart, July 2001). Such a function gives similar or slightly better fits. To keep the analytical derivations in the next chapters transparent, Eq. 3.7 is preferred. For all Zimbabwean and world locations in this dissertation the calibrated versions of Eq. 3.7 for p_{01} and p_{11} only surpassed the probability of 1 at unrealistically high values of the monthly rainfall P_m.[14] In this dissertation Eq. 3.7 is therefore used. In the case of computer models the argument of transparency is less valid. If the equations in this dissertation are implemented in models that use stochastically-generated monthly rainfall data, the use of Eq. 3.8 is preferable to avoid any possibility of probabilities higher than 1.

The transition probability from a dry day to a rain-day, is thus modelled as:

$$\boxed{p_{01} = q(P_m)^r}$$ Eq. 3.9

---

[13] This is a logistic function $y = \dfrac{\theta_3}{1 + \theta_1 \exp(\theta_2 x)}$ with the θ_3 set to 1 and by substitution of x by $\ln(P_m)$.

[14] For Zimbabwean stations the monthly rainfall at which p_{11} surpassed 1 is 818 mm/month, while the historically highest monthly rainfall was less than 500 mm/month.

Box 3-B   Assessing transition probabilities

The estimated probability p_{01} is straightforward:

$$p_{01} = \frac{a}{A}$$

Eq. 3b.1

where

A is the total number of days in the sample preceded by a dry day (-),

a the number of rain-days preceded by a dry day (-)

Imagine a sample 01001001100 (dry, rain, dry, dry...). Then $A = 6$ and $a = 3$ and the estimated probability $p_{01} = 50\%$. Imagine a sample 10100. Then $A = 2$ and $a = 1$. This also has an estimated $p_{01} = 50\%$, but it is obvious that the accuracy of this estimate is lower. Because of the Markov property, the number of days that are preceded by a certain number of successive dry days decreases exponentially with the increasing number of dry days that are preconditional. Thus, the probability of having another rain-day after 3 consecutive rain-days (p_{0001}) can be derived with less accuracy than p_{01}.

When choosing class sizes, the question is: how many months are necessary in a sample to give a reasonably accurate probability of p_{01}? Or: how accurate is the estimate of a transition probability, as a function of a and A?

The Bernoulli distribution describes probabilities of 'success' or 'failure', i.e. 'dry' or 'rain'. These probabilities are not related to previous time steps. Yet, if only the sample of days with the same condition on the previous day is considered, the probability of having rain-days in this sample could be judged in the same way as in a Bernoulli process.

Because there are only two possible states, a day is a dry day or a rain-day, the analysis of the time series to assess p_{01} is a binomial distribution. Thus, the mass function is:

$$f(p_{01} * A) = \frac{A!}{(p_{01} * A)!(A - (p_{01} * A))!} \left(\frac{a}{A}\right)^{a} \left(1 - \frac{a}{A}\right)^{A-a}$$

Eq. 3b.2

This mass function has a standard deviation:

$$\sigma(p_{01}) = \frac{\sqrt{a\left(1 - \frac{a}{A}\right)}}{A}$$

Eq. 3b.3

Eq. 3b.1 and 3b.2 are used to determine the error bars in Figures 3.1, 3.2 and 3.3.

The first n days in a month are not used in the determination of the transition probability of a rain-day after a spell of at least n days. This is considered in the determination of the error bars. Similarly to the above, equations for p_{11} are derived.

A wet spell (or dry spell) of n days occurs 4 times in the sample with a minimum of n-4 preceding days in the spell, 3 times in the sample with a minimum of n-3 preceding days in the spell and so on. This means that neither the variance nor the estimated probabilities are completely independent and that trends can be spurious.

Through calibration,[15] coefficients for the different stations with daily data have been derived:

$$p_{01\,har} = 0.020\,P_m^{0.55}$$

$$p_{01\,mas} = 0.030\,P_m^{0.43} \qquad \text{Eq. 3.10}$$

$$p_{01\,bul} = 0.044\,P_m^{0.34}$$

See Figure 3.5. Only for very high monthly rainfall in Bulawayo does the probability derived from the observations deviate substantially from the fitted equations, although in that case the model fits better to the scatter plot of individual months. This may indicate that months with extremely high rainfall in Bulawayo (in the south of Zimbabwe) are the result of a different climatic situation. The differences between the locations are shown in Figure 3.7.

The relationship between monthly rainfall and p_{11} appears to fit the same power function for Harare, Masvingo and Bulawayo. Calibration yields

$$\boxed{p_{11} = u(P_m)^v}$$
$$= 0.20\,P_m^{0.24} \qquad \text{Eq. 3.11}$$

See Figure 3.6.

Subtracting Eq. 3.10 from Eq. 3.11 yields the following expression for C

$$
\begin{aligned}
C &= P_m^{\,v}(u - q\,P_m^{\,r-v}) \\
C_{har} &= P_m^{0.24}(0.20 - 0.020\,P_m^{0.31}) \\
C_{mas} &= P_m^{0.24}(0.20 - 0.030\,P_m^{0.19}) \\
C_{bul} &= P_m^{0.24}(0.20 - 0.044\,P_m^{0.10})
\end{aligned}
\qquad \text{Eq. 3.12}
$$

See Figure 3.8.

For Harare C appears to be almost constant for P_m greater than (say) 50 mm/month, namely 0.34. As we will see later, below 50 mm/month the rainfall is completely intercepted. Therefore, assuming a constant C would be satisfactory in the case of Harare. But for other stations equations of the type of Eq. 3.12 may have to be used.

---

[15] The calibration was done by minimising square errors of the points determined for each class of monthly rainfall.

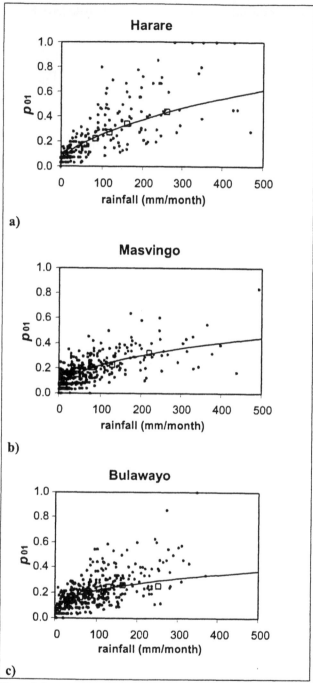

Figure 3.5   Transition probability p_{01} derived from historical data for 1, 2, 3, 4 preceding dry days for (a) Harare, (b) Masvingo, (c) Bulawayo and fitted power function (solid line). The dots (•) are estimates of the transition probabilities determined from the individual months. The squares (□) are estimates of the transition probabilities determined by considering all months in a monthly rainfall class as one sample.

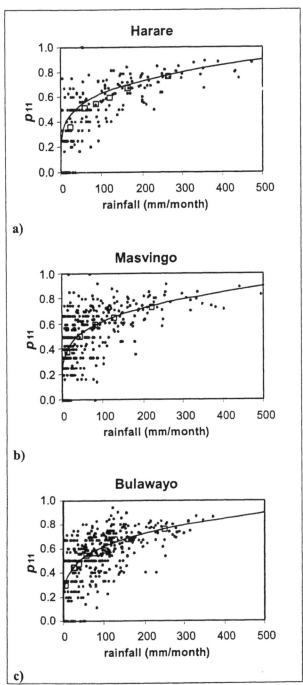

Figure 3.6   Transition   probability p_{11} derived   from historical data for 1, 2, 3, 4 preceding rain-days for (a) Harare, (b) Masvingo, (c) Bulawayo and model $p_{11} = 0.20 \ P_m^{0.24}$. For further explanation, see caption of Figure 3.5.

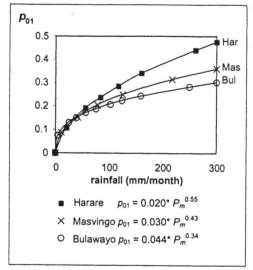

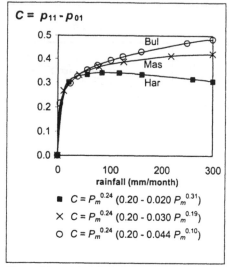

Figure 3.7 Modelled p_{01} in relation to monthly rainfall for Harare, Masvingo, Bulawayo.

Figure 3.8 Modelled difference between p_{11} and p_{01}, C, for Harare, Masvingo, Bulawayo.

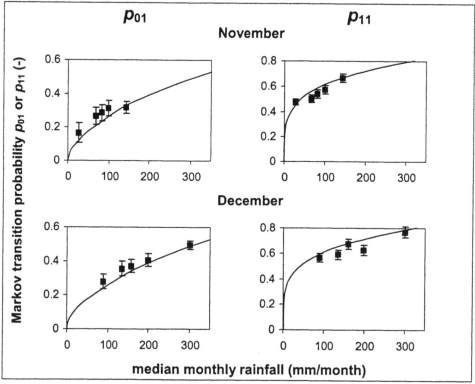

Figure 3.9 Harare Belvedere. For each month in the rainy season the median monthly rainfall in five classes of equal size is plotted against the empirically-derived transition probability p_{01} (left) and p_{11} (right). The drawn line gives the power function that was fitted to data from Harare airport without distinction between seasons. November, December.

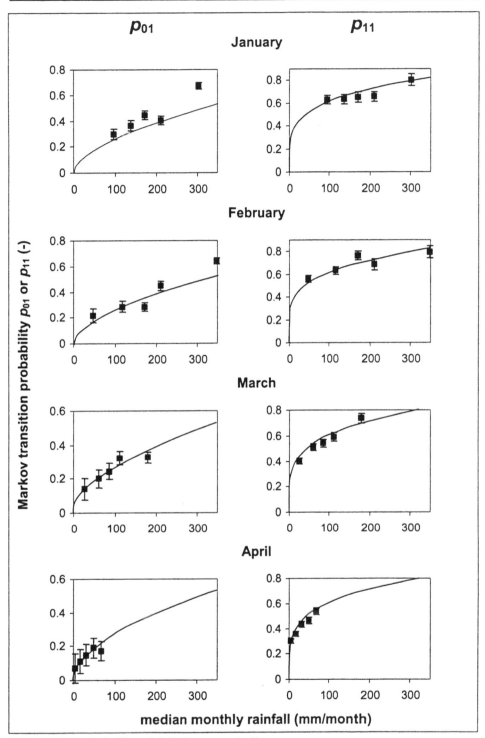

Figure 3.10   As in Figure 3.9, for January to April.

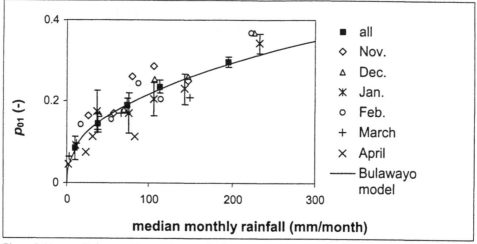

Figure 3.11   Bulawayo Goetz. Scatter plot between median monthly rainfall and p_{01}'s. The black squares (■) give the probability p_{01} where no distinction between months is made. In this case and for January error bars of the standard deviation are shown. Each dot represents thirteen historical months.

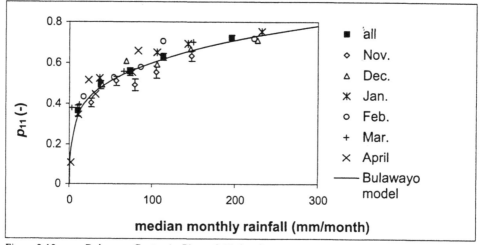

Figure 3.12   Bulawayo Goetz. As Figure 3.11, but for p_{11}.

3.6   Seasonal differences

Section 2.3 presented an overview of stochastic daily rainfall models based on the Markov process. Most of these models used transition probabilities that depend on the time of year. Different months with the same amount of rainfall can show different daily rainfall patterns, as rain-producing mechanisms vary over the season. Different parameters could be calibrated for the different months in a year. However, the accuracy of the transition probability that is derived from historical measurements decreases rapidly with the sample size. To derive transition probabilities for each of the six months in the rainy season separately would require about six times more data (see Section 2.3 and box 3-B).

For the station Harare Belvedere daily data are available for the period 1936-1994.[16] This makes it suitable to test the hypothesis of no dependency of the transition probabilities on the month in the season. Figures 3.9 and 3.10 show that the power relations between monthly rainfall and the transition probabilities that were derived for Harare airport without distinction of season, fit well for all months.

For Bulawayo the results are shown in a more compressed form in Figures 3.11 and 3.12. November and April, the start and end of the rainy season, show the largest deviations from the overall fitted power relation. The data points of p_{01} are greater at the start of the rainy season than at the end, as a function of the monthly rainfall. Calibrating for each month separately explains little of the variance.

It is concluded that it is not necessary to make a distinction between months in the rainy season, at least for Zimbabwe. In Section 3.8 it will be shown that on the basis of stochastically generated time series, this conclusion is not valid for all locations in the world.

3.7 Spatial differences

Stern & Coe (1984) used Markov processes for daily rainfall in analytical models to describe useful parameters for agriculture. For example, they derived an expression for the most probable start of the rainy season. However, they suggested that for spatial interpolation of the parameters that describe the seasonal variation of the transition probabilities, Tanzania would need good quality daily rainfall records 1,500 rain stations (comment by R. Mead on the verbal presentation of the paper by Stern & Coe, 1984, same edition). This makes their method unsuitable for large-scale application. Jones & Thornton (1997), however, compared parameter fits on third order autoregressive processes for 3,000 stations all over the world and concluded that (i) similar climates from different parts of the globe result in characteristic model parameters; and (ii) within a climate type, there are broad similarities in model parameters.

It is the objective of this dissertation to develop a method that can be calibrated with daily data from a few well separated locations (~300 km). In this Section data from five other Zimbabwean stations are presented. The locations are shown in Figure 3.13 and calibrated models of p_{01} and p_{11} are depicted in Figure 3.14. The models are calibrated by minimisation of the sum of squares of the deviations for the Markov transition probabilities derived from all months in classes of monthly rainfall.

Chinhoyi (825 mm/y, 1,145 m alt.) lies some 100 km northwest of Harare and at an altitude of about 300 m less than Harare (1,470 m alt.). The mean annual rainfall is similar to that of Harare. It appears that the transition probability p_{01} is higher than that for Harare, suggesting more wet spells.

---

[16] These data only became available after the analyses based on Harare airport data (1959-1993).

Figure 3.13   Map of Zimbabwe with stations used in this
Section.

Mutare (750 mm/y, 1,110 m alt.) is at about the same altitude as Chinhoyi, but it is
surrounded by the Eastern Highlands. Values of p_{01} are visibly smaller than for
Harare.

Rusape (810 mm/y, 1,430 m alt.) is situated about 75 km from Mutare about one third
of the distance between Mutare and Harare. Rusape lies higher than its surroundings.
Despite the fact that altitude and mean annual rainfall are similar to those of Harare,
the calibrated model for p_{01} is similar to that for Mutare. The difference with Mutare
is not visible. Thus, interpolation between Harare and Mutare gives a worse result
than using the parameters for Mutare.

Kwekwe (655 mm/y, 1,200 m) lies halfway between Harare and Bulawayo (630
mm/y, 1,330 m alt.). All three stations are located on the highveld, but Kwekwe has
the lowest altitude. The calibrated model of p_{01} for Kwekwe is a normal average of
the models for Harare and Bulawayo. However, the line marked 'interpolate' shows
that averaging the parameters in the power functions of Harare and Bulawayo does
not fit. This is logical, because the relations are not linear.

Zvishavane (600 mm/y, 915 m alt.) is at about one third distance between Masvingo
(640 mm/y, 1,100 m alt.) and Bulawayo, at a lower altitude. Despite nearness and
different altitude, the calibrated power function for p_{01} resembles that of Bulawayo
more than that for Masvingo.

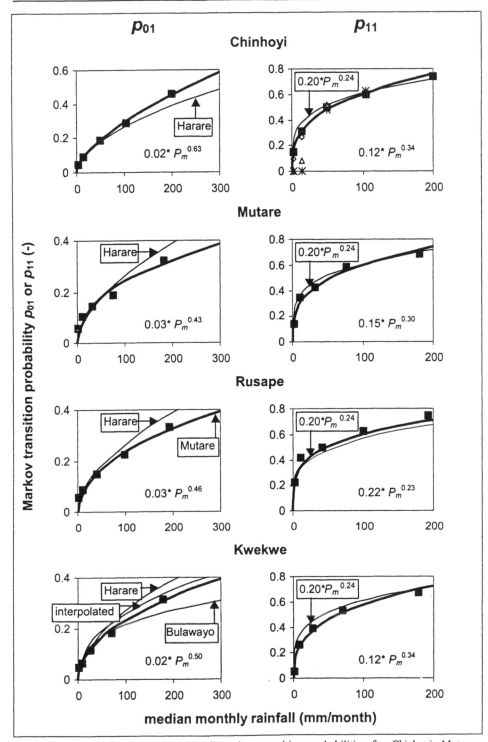

Figure 3.14   Plots of monthly rainfall against transition probabilities for Chinhoyi, Mutare, Rusape, Kwekwe. For explanation see Figure 3.15.

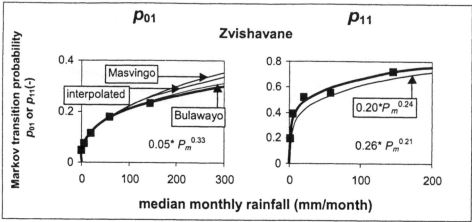

Figure 3.15   Plots of monthly rainfall against transition probabilities for Zvishavane. In each chart the bold line is the model fitted to the transition probabilities (■). The fitted model is annotated in the right lower corner of the graph. For p_{01} the models of the nearest stations Harare, Masvingo and/or Bulawayo are also shown. For p_{11} the model that is valid in Harare, Masvingo and Bulawayo, $p_{11}=0.20*P_m^{0.24}$, is shown for reference. In the chart p_{01} of Kwekwe and Zvishavane, the line marked 'interpolation' shows the model that is found by averaging the parameters in the power function for Harare and Bulawayo, and Masvingo and Bulawayo, respectively. The averaging is weighted with the inverse distance.

Among the stations presented, Chinhoyi has the highest and Bulawayo the lowest estimates of p_{01}, respectively around 0.6 and 0.3 at $P_m = 300$ mm/month. For p_{11} the power function $p_{11} = 0.20*P_m^{0.24}$ is adequate for all locations. Zvishavane shows the largest deviations. All the stations presented here are at relatively high altitude. No stations in the Lowveld or in the Zambezi Valley are represented, due to the fact that the daily data from these stations are not yet digitised. Extrapolation may possibly be wrong.

Conclusions:
- The spatial variability in the power function that describes the relationship between monthly rainfall and the transition probability p_{11} is very low.
- For p_{01} the relationships are more variable, but they are generally between the relations fitted to stations at some 300 km distance. Study of a few locations in Zimbabwe showed that the relationship in some cases can best be interpolated by weighting with the inverse of the distance (Kwekwe), while in other cases the relationship of the nearest station gives the best results (Rusape). However, an example is also given of a station that has a similar relationship to that of a station that is neither the nearest nor the one most similar in altitude (Zvishavane). Although topography is widely used as a covariant with rainfall to improve interpolation (Blöschl & Sivapalan, 1995) and this covariance is recognised in Zimbabwe (Torrance, 1981), the analysis above does not indicate the usefulness of interpolation employing covariance with topographical data.
- Because the relations are not linear, a weighted average of the parameters that describe the power function does not yield the relationship that is looked for. Thus, for a location z between two locations x and y,

$$p_{01z} = \frac{p_{01x} + a*p_{01y}}{1+a} \neq \left(\frac{q_x + a*q_y}{1+a}\right) P_m^{\left(\frac{r_x + a*r_y}{1+a}\right)}.$$

The weighting process has to be further explored, but the best option currently is to use the relationship of the nearest station for which daily records are available, unless more knowledge about the climate is available.

3.8   Other locations in the world

Zimbabwe is used here as a case study for a generic problem. The models that are presented in the next chapters are based on power relations between the monthly rainfall and the transition probabilities p_{01} and p_{11}. It was tested if such power functions are valid worldwide, at locations for which Markov models for daily rainfall have proven applicable.

For this purpose, artificial time series generated with the Markov models as they are described in the literature have been used. All these models used harmonics to describe the seasonal variation of the transition probabilities and of the parameters of the probability density functions for rain on rain-days:

- Zucchini et al. (1992) has set out the parameters for Peters Gate, which lies close to Cape Town in South Africa. Rainfall on rain-days is described with a Weibull distribution.
- Stern & Coe (1982) depicted Hyderabad in western India. The model is for the rainy season only, from April until October. Rainfall on rain-days is described with a gamma distribution. The probability density function of rain on rain-days is different depending on whether the previous day is a dry or a rain-day.[17]
- Woolhiser & Pegram (1979) made a Markov model for four stations in the USA: Indianapolis in the state of Indiana, Kansas City in Missouri, Sheridan in Wyoming and Tallahassee in Florida. They used 20-25 years of historical data for each station. They selected these stations to include substantially different climatic conditions. Rainfall on rain-days is described with a mixed exponential distribution.

For each station 40 years of artificial rainfall series were generated.

Figures 3.16 and 3.17 show the results for different stations. Table 3-2 summarises the calibrated parameters. The figures confirm that the relations between the monthly rainfall and the transition probabilities p_{01} and p_{11} are power functions. The models that are shown in the next chapters are applicable for any value for the parameters in the power functions and can also be applied with logistic functions as in Eq. 3.8.

---

[17] In the paper by Stern & Coe (1982) harmonic series are described for the log of the mean rain per rain-day: log (μ_{Pr}). However, with the given parameters this does not yield the mean rainfall that is depicted in figures in the same paper. If log(μ_{Pr}) is replaced by ln(μ_{Pr}), results are more similar. Therefore, ln(μ_{Pr}) is used here.

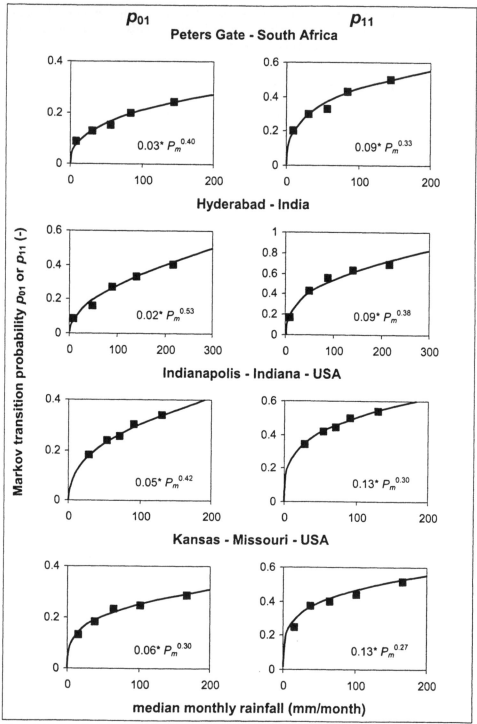

Figure 3.16   Relations between monthly rainfall and transition probabilities p_{01} and p_{11}, derived from artificial rainfall series for Peters Gate, Hyderabad, Indianapolis and Kansas.

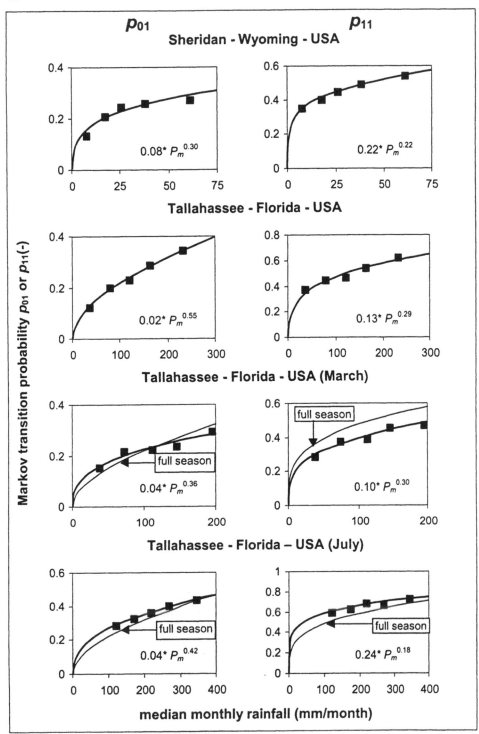

Figure 3.17   Continuation of Figure 3.16, for Sheridan and Tallahassee. The Tallahassee charts show the full season, March and July.

In the Zimbabwean case it has been assumed that a distinction between months is not necessary. Tallahassee in Florida, however, shows an extreme seasonal variability in Markov transition probabilities. Between March and July, p_{01} varies between 0.2 and 0.35 and p_{11} between 0.42 and 0.65.[18] Artificial rainfall time series for March and July were generated (40*12 months). The figures show that there is a distinct difference between the models calibrated on the individual months and on the full season. However, it needs to be noted that this may be inherent in the use of artificially generated time series and does not necessarily reflect the historical data. The Markov stochastic rainfall models use transition probabilities that are calibrated to reflect the most probable rainfall occurrence in a certain season. These methods do not necessarily preserve the variances of monthly precipitation (Buishand, 1977; Zucchini & Adamson, 1984) and therefore neither do they preserve the correct variance in transition probabilities for a particular monthly rainfall. If sufficient historical data are available, there is no problem calibrating different power relations for different months and using these calibrated parameters in the monthly models described in the following chapters.

Table 3-2   Calibrated model parameters for different rain stations.

Location	$p_{01} = q\,P_m^{\,r}$		$p_{11} = u\,P_m^{\,v}$	
	q	**r**	**u**	**v**
Harare	0.020	0.55	0.20	0.24
Masvingo	0.030	0.43	0.20	0.24
Bulawayo	0.044	0.34	0.20	0.24
Peters Gate (SA)	0.094	0.33	0.034	0.40
Hyderabad (India)	0.092	0.38	0.024	0.53
Indianapolis (Indiana, USA)	0.129	0.30	0.045	0.42
Kansas (Missouri, USA)	0.129	0.27	0.061	0.30
Sheridan (Wyoming, USA)	0.216	0.22	0.084	0.30
Tallahassee (Florida, US)	0.127	0.29	0.017	0.55

3.9   Conclusions

In this Chapter the following conclusions have been drawn:

- Two state Markov models can be applied to describe daily variability of rainfall occurrence in Zimbabwe.

- The transition probabilities of rainfall occurrence after a dry day p_{01} and rainfall occurrence after a rain-day p_{11} can be expressed as power functions of monthly rainfall ($p_{01} = qP_m^{\,r}$, $p_{11} = uP_m^{\,v}$). This is applicable not only to Zimbabwe, but also to other locations in the world for which Markov rainfall models have been published (South Africa, India, USA).

- Instead of power functions, logistic functions of the form $1/(1+\theta_l P_m^{\,\theta_2})$ are suitable. Such functions are theoretically more correct, because the value 1 cannot be surpassed. However, power functions are more transparent and the fitted power functions only surpass 1 at unrealistically high values of monthly rainfall. Therefore, in this dissertation power functions are used.

---

[18] The harmonics model of Woolhiser & Pegram (1979) yields a p_{01} of about 0.35 for July, but the historically-derived p_{01}, as depicted in the same paper, reaches about 0.5.

- For Zimbabwe (Harare, Masvingo and Bulawayo) the transition probabilities are not dependent on the month of the rainy season. However, this cannot be applied generally to all locations worldwide.

- For Zimbabwe, the spatial variability of the power functions that describe the relationships between monthly rainfall and either of the two transition probabilities is small. For the transition probability of a rain-day after a rain-day p_{11} the same parameters were calibrated for all Zimbabwean stations. Rain stations at a spatial scale in the order of 300 km are sufficient to get reasonably good fits for the relationship between monthly rainfall and the probability of a rain-day after a dry day p_{01}.

4 Monthly Parameters Representing Rainfall Occurrence

4.1 Introduction

In this Chapter the relationship between the transition probabilities of the two state Markov chain and parameters for the frequency and clustering of rain-days are derived. Probability density functions are presented for the number of rain-days per month, the number of wet spells and the number of days in dry and wet spells.

The overall objective is to find relationships between monthly water fluxes for locations near rain stations that do not have daily rain data available. To date, relationships between monthly rainfall and moisture fluxes have been derived from daily rain data. Parameters that describe such empirical relations may be interpolated between different locations. However, spatial interpolation of the parameters describing the Markov property as a function of monthly rainfall appears a more sensible approach, because:

- Relationships between rainfall and moisture fluxes depend on physical characteristics of the soil and the vegetation. These are very inhomogeneous in space and hence difficult to interpolate. The statistical characteristics of rainfall are related to the dominant rainfall-producing processes, which are far more homogeneous.
- Fewer parameters need be interpolated.

4.2 Number of rain-days

Gabriel & Neumann (1962) arrived at an exact solution for the probability of n_r rain-days in any n days, via steps described in Box 4-A. This relationship is:

$$P(n_r|n) = \frac{p_{01}}{1-C} P(n_r|n,1) + \left(1 - \frac{p_{01}}{1-C}\right) P(n_r|n,0) \qquad \text{Eq. 4.1}$$

where $P(n_r|n,1)$ is the probability of exactly n_r rain-days within a period of n days following a rain-day and $P(n_r|n,0)$ is the probability of exactly n_r rain-days within a period of n days following a dry day (see Box 4-A). $C = p_{11} + p_{01}$, see Eq. 3.5.

For a large value of n the distribution of n_r tends to a normal distribution with mean

$$E(n_r|n) = n\frac{p_{01}}{1-C} \qquad (-) \qquad \text{Eq. 4.2}$$

and variance

$$\text{Var}(n_r|n) = n\frac{p_{01}}{1-C}\left(1 - \frac{p_{01}}{1-C}\right)\frac{1+C}{1-C} \qquad (-) \qquad \text{Eq. 4.3}$$

(Gabriel & Neumann, 1962) In this context n and n_r are counters of days, thus they are dimensionless.

Box 4-A   Probabilities of n_r rain-days among n days
(after Gabriel & Neumann, 1962)

When using a two state Markov process, the probability of exactly n_r rain-days among n days following a rain-day is:

$$P(n_r|n,1) = p_{11}{}^{n_r}(1-p_{01})^{n-n_r}\sum_{u=1}^{u_1}\binom{n_r}{v}\binom{n-n_r-1}{w-1}\left(\frac{1-p_{11}}{1-p_{01}}\right)^w\left(\frac{p_{01}}{p_{11}}\right)^v \qquad \text{Eq. 4a.1}$$

where

$$u_1 = n + \frac{1}{2} - \left|2n_r - n + \frac{1}{2}\right| \qquad\qquad \text{if } n_r < n \qquad \text{Eq. 4a.2}$$

and where v is the lowest integer not smaller than a $1/2*(u-1)$

$$v = \text{Int}\left(\frac{u}{2}\right) \qquad\qquad \text{Eq. 4a.3}$$

and where w is the lowest integer not smaller than $1/2*u$

$$w = \text{Int}\left(\frac{u+1}{2}\right) \qquad\qquad \text{if } n_r < n \qquad \text{Eq. 4a.4}$$

The probability that after a rain-day all n days are rain-days is:

$$P(n_r|n,1) = p_{11}{}^{n_r} \qquad\qquad \text{if } n_r = n \qquad \text{Eq. 4a.5}$$

The probability of exactly n_r rain-days among n days following a dry day is symmetrical. If $n_r > 0$:

$$P(n_r|n,0) = p_{11}{}^{n_r}(1-p_{01})^{n-n_r}\sum_{u=1}^{u_1}\binom{n_r-1}{w-1}\binom{n-n_r}{v}\left(\frac{1-p_{11}}{1-p_{01}}\right)^v\left(\frac{p_{01}}{p_{11}}\right)^w \qquad \text{Eq. 4a.6}$$

where

$$u_1 = n + \frac{1}{2} - \left|2n_r - n - \frac{1}{2}\right| \qquad\qquad \text{if } n_r > 0 \qquad \text{Eq. 4a.7}$$

and

$$P(n_r|n,1) = p_{00}{}^{n} \qquad\qquad \text{if } n_r = 0 \qquad \text{Eq. 4a.8}$$

All of these transition probabilities Eqs. 4a.1, 4a.5, 4a.6 and 4a.8 together determine the probability of n_r rain-days among any n days. The solution is given in the main text.

Katz (1974) derived the above relations using transition matrices, see Appendix A.

"The correlation between numbers of rain-days in successive periods of n_x days results entirely From the dependence of the event on the first day of the second period and that on the last day of the first period. Clearly the correlation will be small for large n. Thus, successive long periods will have virtually independent numbers of rain-days." (Gabriel & Neumann, 1962). It should be realised that $\text{Var}(n_r)$ is in (days/month)^2, while substitution of the units on the right-hand side of the equal sign gives (days/month). Thus, unit wise, Eq. 4.3 is not correct. The n and n_r should be regarded as unitless counters.

Figure 4.1a and b depict the exact solution given by as a function of n_r for $n = 30$ days. The approximation of the expected number of rain-days, given in Eq. 4.2, deviates 1% in the case of $p_{01} = 0.3$ and $p_{11} = 0.7$ (Figure 4.1a) and 11% in the unrealistic case that $p_{01} = 0.1$ and $p_{11} = 0.9$ (Figure 4.1b).

Thus, a satisfactory approximation for the average number of rain-days n_r in a month with n_m days (-) is

$$E(n_r|n_m) = n_m \frac{p_{01}}{1-C}$$

(-)   Eq. 4.4

with variance

$$Var(n_r|n_m) = n_m \frac{p_{01}}{1-C}\left(1 - \frac{p_{01}}{1-C}\right)\frac{1+C}{1-C}$$

(-)   Eq. 4.5

In these p_{01} and C can be expressed as a function of monthly rainfall P_m (Eqs. 3.10 and 3.12). In Figure 4.2 the resulting models for Harare, Masvingo and Bulawayo are shown and compared with historical data. The figure shows that Eqs. 4.4 and 4.5 give a good representation of the expected number of rain-days for a certain monthly rainfall.

For further derivations, an expression for the probability of a rain-day is needed. This follows directly from Eq. 4.4:

$$p = P(P_d > 0) = \frac{E(n_r|n_m)}{n_m} = \frac{p_{01}}{1-C}$$

Eq. 4.6

For further reference, Clarke (1998) presents a lucid summary of the methods described by Stern & Coe (1984).

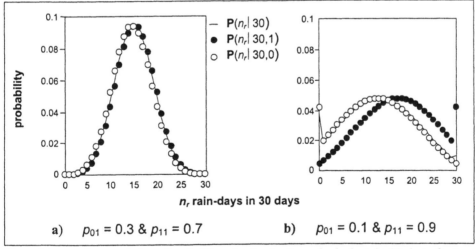

Figure 4.1   Exact solution of probability density function of rain-days in 30 days, for two different combinations of p_{01} and p_{11} (see Box 4-A). The symbol ● gives the density function $P(n_r|n,1)$ for 30 days following a rain-day, the symbol O is $P(n_r|n,0)$ for 30 days following a dry day and the drawn line is an average between the two density functions weighted for the probability that the day before the 30 days was a dry or a rain-day.

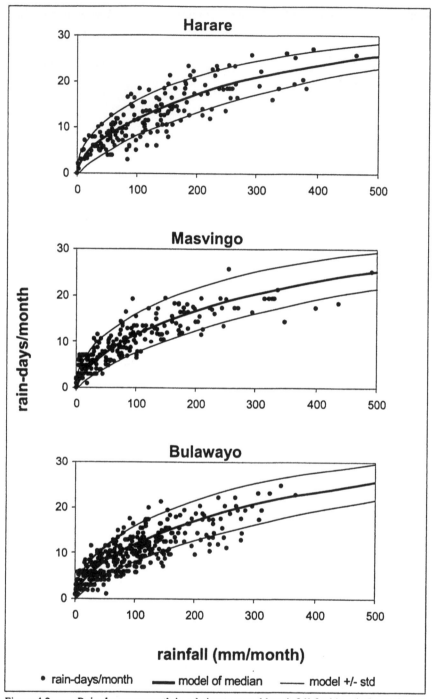

Figure 4.2   Rain-days per month in relation to monthly rainfall for historical months and for model based on Markov process. The band around the model gives the theoretical standard deviation. In the model a month is assumed to have 30 days. The historical number of rain-days is multiplied with $30/n_m$ to correct for this.

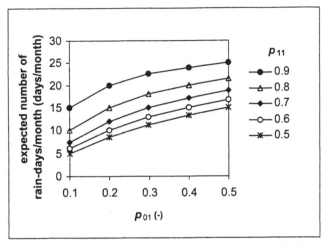

Figure 4.3   Expected number of rain-days in month (30 days) for different values of p_{01} and p_{11}.

4.3   Lengths of dry spells

In the daily practice of the hydrologist and agricultural engineer in a country such as Zimbabwe, the length of dry spells is commonly used. During a dry spell transpiration slowly depletes the soil moisture. The frequency distribution of the length of dry spells, for example, is important for design capacities of supplementary irrigation works.

Gabriel & Neumann (1962) obtained the probability distribution for lengths of different dry spells. A dry spell of length n has a probability density function:

$$P(n_{dry} = n) = p_{01} * (p_{00})^{n-1}$$ Eq. 4.7

Since all equations are presented as functions of p_{11} or p_{01} and bearing in mind that $p_{01} + p_{00} = 1$, the equations for dry spells can simply be written in the following manner,

$$P(n_{dry} = n) = p_{01} * (1 - p_{01})^{n-1}$$ Eq. 4.8

and

$$P(n_{dry} \geq n) = (1 - p_{01})^{n-1}$$ Eq. 4.9

from which follows

$$P(n_{dry} > n) = (1 - p_{01})^{n}$$ Eq. 4.10

Eq. 4.9 will give as the first moment the expected number of days in a dry spell.[19]

$$E(n_{dry}) = \frac{1}{p_{01}} \Delta t_d \qquad \text{(days)} \qquad \text{Eq. 4.11}$$

with a variance

$$Var(n_{dry}) = \frac{1 - p_{01}}{(p_{01})^2} (\Delta t_d)^2 \qquad \text{(days)}^2 \qquad \text{Eq. 4.12}$$

where Δt_d is 1 day and is introduced for consistency in units.

In Figure 4.4 the measured probability density functions of the spell lengths are presented for different class intervals of monthly rainfall, for Harare, Masvingo and Bulawayo. The model of Eq. 4.8 has been applied on the median of the sample in each class. Only spells were taken into account that did not cross the month's boundaries. This particularly cuts out long dry spells and therefore may explain why the model gives a higher probability of long dry spells. However, an approach that allocated the spells that crossed a months boundary to the month at the time of the middle of the dry spell gave a larger deviation between model and data (not shown).

Jimoh & Webster (1996) suggested fitting on the histograms of lengths of wet spells. Figure 4.5 shows that other power functions for p_{01} can be found for which the theoretical probability density functions of lengths of dry spells (Eq. 4.11) fit better to the histograms.

Figure 4.6 shows the average dry spell length in relation to the monthly rainfall, measured and modelled, with the original transition probabilities. The model overestimates average dry spell lengths in all locations, in particular in months of little monthly rainfall.

---

[19] Statisticians are unfamiliar with hydrological terms such as dry and wet spells. In statistical textbooks, e.g. Grimmett & Stirzaker (2001), Markov processes are analysed for their **recurrence** times. The **recurrence time** is the time elapsed between two events, in the case of a dry spell the two events are rain-days. If the recurrence time $t_{11} = 1$, the rain-day is immediately followed by a rain-day, thus there is no dry spell. If the recurrence time $t_{11} = 2$, the dry spell is only one day long. For transfers between statistical textbooks and hydrological applications, comparison of first moments shows that

$$E(n_{dry}) = \frac{E(t_{11}) - 1}{1 - p_{11}}.$$

Foufoula-Georgiou & Lettenmaier (1987), who developed stochastic rainfall models (see Chapter 2), call the recurrence time the interarrival time.

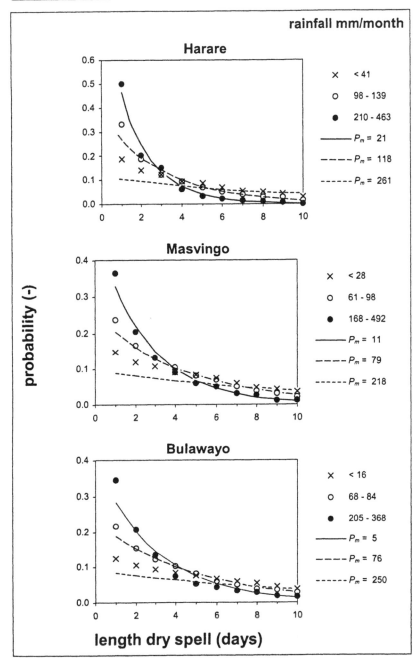

Figure 4.4   Probability density functions of dry spell lengths, measured and modelled for Harare, Masvingo, Bulawayo. The models use the median of the monthly rainfall in the sample of months in each class. Each class has the same number of months in the sample (about 40). Spells that crossed the boundaries of months are not taken into account.

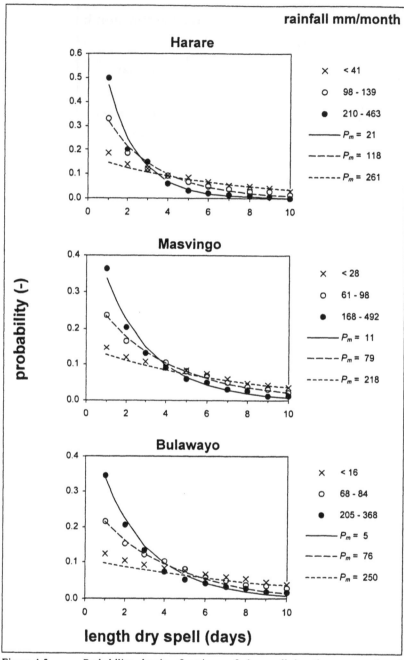

Figure 4.5   Probability density functions of dry spell lengths, measured and modelled for Harare, Masvingo, Bulawayo. Contrary to the models of Figure 4.4, the power functions that describe the relationship between monthly rainfall and transition probabilities were fitted to minimise the sum of the squares of the differences between the measurements and the models for all rainfall groups. For the models the median of the monthly rainfall in the group of measurements was used. For Harare p_{01}= $0.04*P_m^{0.44}$ instead of $p_{01} = 0.020*P_m^{0.56}$, for Masvingo $p_{01} = 0.058*P_m^{0.32}$ instead of p_{01} $= 0.030*P_m^{0.43}$ and for Bulawayo $p_{01} = 0.054*P_m^{0.33}$ instead of $p_{01} = 0.044*P_m^{0.34}$.

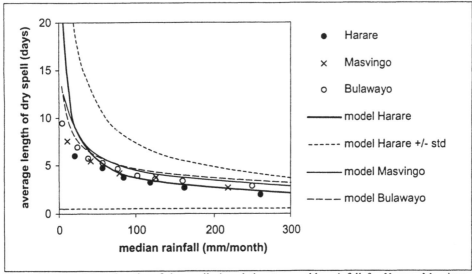

Figure 4.6   Mean lengths of dry spells in relation to monthly rainfall for Harare, Masvingo, Bulawayo, observed (markers) and theoretical (lines). For Harare the means plus/minus the theoretical standard deviation in length of dry spell is given also.

4.4   Lengths of wet spells

Obviously, the statistics of the lengths of wet spells directly follow from the equations presented for dry spells, by replacing the transition probability of a dry day after a dry day, p_{00}, by the transition probability of a rain-day succeeding a rain-day, p_{11}. The probability of lengths of wet spells n_w therefore reads (Şen, 1976):

$$\mathbf{P}(n_{wet} = n) = (1 - p_{11}) * p_{11}^{n-1}$$

Eq. 4.13

and

$$\mathbf{P}(n_{wet} \geq n) = p_{11}^{n-1}$$

Eq. 4.14

from which follows

$$\mathbf{P}(n_{wet} > n) = p_{11}^{n}$$

Eq. 4.15

See Figure 4.8 for the probability density function. The difference between Eqs. 4.14 and 4.15 is subtle, but important.

In Figure 4.7 the observed probability density functions of the different classes are given, for Harare, Masvingo and Bulawayo. For the lowest, the middle and the highest class the theoretical probability density functions are also given, based on the transition probabilities as derived for the median rainfall in the class. Determining the histograms of the length of wet spells causes less severe problems as for dry spells, because wet spells are generally shorter than dry spells and therefore the boundaries of months are not so influential. Expressing the probability p_{11} as a function of monthly rainfall, in the way that was derived in Section 3.5, gives a good fit with the observed probability density function.

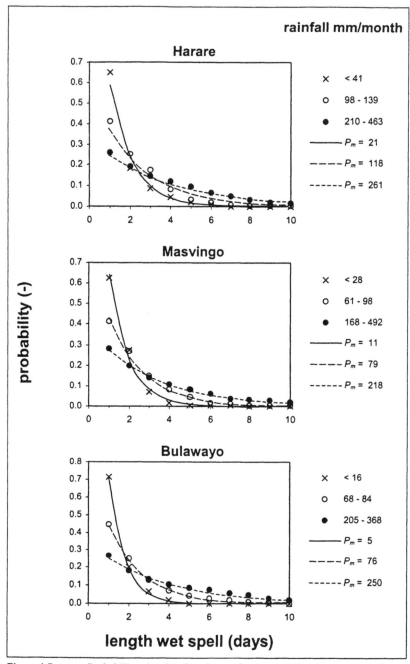

Figure 4.7   Probability density functions of wet spell lengths, measured and modelled for Harare, Masvingo and Bulawayo. The model uses the median of the monthly rainfall in a sample of months in each class. Each class has the same number of months in the sample (about 40). Spells that crossed the boundaries of months are not taken into account.

The first moment of the density function is the expected number of rain-days in a wet spell:

$$\mu_{n_{wet}} = E(n_{wet}) = \frac{1}{(1 - p_{11})} \Delta t_d \qquad \text{(days)} \qquad \text{Eq. 4.16}$$

The spell length has a variance that is

$$\sigma^2_{n_{wet}} = \text{Var}(n_{wet}) = \frac{p_{11}}{(1 - p_{11})^2} (\Delta t_d)^2 \qquad \text{(days)}^2 \qquad \text{Eq. 4.17}$$

In Figure 4.10 the expected length of a wet spell in relation to the monthly rainfall is given, for all classes of the historical series and for the theoretical model. The model performs remarkably well, given the standard deviation in length of wet spell. The observed values are slightly lower than the modelled ones. This may be partly due to the boundaries of months, in the same way as is explained for the dry spells.

Sharma (1996a) needed $P(n_{wet} > n)$ in equations for the design of rain-water catchment systems.

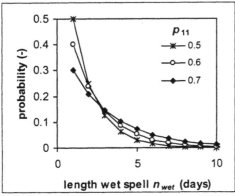

Figure 4.8   Probability density function of length of wet spell.

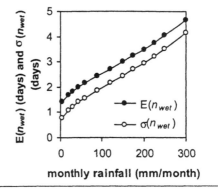

Figure 4.9   Theoretical expectation and standard deviation of lengths of wet spells in relation to monthly rainfall. Expectation and standard deviation have almost the same slope.

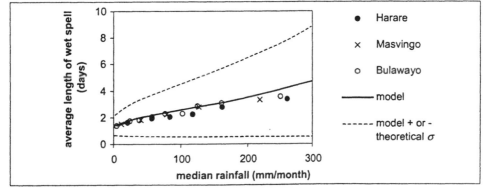

Figure 4.10   The theoretical and observed mean lengths of wet spells in relation to monthly rainfall. The dotted lines are the theoretical mean lengths + or - the standard deviation in the lengths of the wet spells. Since p_{11} was calibrated the same for Harare, Masvingo and Bulawayo, there is no difference in model between the locations.

4.5   The number of spells

Sharma (1996 a,b) presents a relationship (derived by Şen, 1980) for the number of wet spells N_{wet} in a period of n_x days. The relationship is a Poisson distribution, which for a month, where $n_x = n_m$, yields

$$P\langle N_{wet} | n_m \rangle = \frac{\exp(-\lambda)*(\lambda)^{N_{wet}}}{N_{wet}!} \qquad \text{Eq. 4.18}$$

with

$$\lambda_{wet} = p(1-p_{11})n_m \qquad (\text{-}) \qquad \text{Eq. 4.19}$$

where n_m is the counter of the number of days in a month and where p is the probability of occurrence of a rain-day (Eq. 4.6).

Because a spell cannot be shorter than one day and has to be surrounded by spells of the opposite type, the number of wet spells in a month with rainfall is within the range:

$$0 < N_{wet} \leq \text{trunc}\left(\frac{n_m - 1}{2}\right) + 1 < \frac{n_m}{2} + 1 \qquad N_{wet} \in \mathbb{N} \qquad \text{Eq. 4.20}$$

In a month of 31 days, the upper limit of the number of wet spells is 16. For 30 or 29 days, the upper limit is 15; for 28 days it is 14. More important is the fact that N_{wet} can not be zero. This limitation to N_{wet} is considerably affecting the Poisson distribution for values of $\lambda < 4$, which in Zimbabwe is valid for months with rainfall less than 100 mm/month. Therefore the Poisson distribution needs to be corrected to become a suitable probability density function for the number of wet spells in a month.[20]

$$P\langle N_{wet} | n_m \rangle = \frac{(\lambda)^{N_{wet}}}{N_{wet}!} * \frac{\exp(-\lambda)}{1-\exp(-\lambda)} \qquad \text{for } 0 < N_{wet} < 17 \qquad \text{Eq. 4.21}$$

---

[20] If Eq. 4.18 were to be used, the sum of the probabilities of all possible lengths of wet spells is less than 1. Eq. 4.18 should thus be corrected to:

$$P\langle N_{wet} | n_m \rangle = \frac{\exp(-\lambda)*(\lambda)^{N_{wet}}}{N_{wet}!} * \frac{1}{\exp(-\lambda)\left(\lambda + \frac{\lambda^2}{2!} + \frac{\lambda^3}{3!} ... + \frac{\lambda^{16}}{16!}\right)}$$

The standard Taylor development for $\exp(\lambda)$ reads (e.g. Almering et al., 1988):

$$\exp(\lambda) = 1 + \lambda + \frac{\lambda^2}{2!} + ... + \frac{\lambda^n}{n!} + O(\lambda^{n+1})$$

The upper limit of the number of wet spells is not important; $O(\lambda^{15}) < 0.001$. Combining the two equations, the probability density function for the number of wet spells becomes Eq. 4.20. For months with little rainfall, the correction factor is considerable. For the Zimbabwean stations with rainfall of 20 mm/month it is around 1.2. For monthly rainfall > 50 mm/month the correction factor is smaller than 1.01.

As for wet spells, a Poisson distribution can be derived for dry spells. For dry spells, the transition probability of having a rain-day in Eq. 4.19, should be replaced by that of having a dry day $(1-p)$. Similarly, the transition probability of a rain-day after a rain-day (p_{11}) should be replaced by that of having a dry day after a dry day $(p_{00} = 1 - p_{01})$:

$$P\langle N_{dry} | n_m \rangle = \frac{\exp(-\lambda_{dry}) * (\lambda_{dry})^{N_{dry}}}{N_{dry}!}$$
$$\text{Eq. 4.22}$$

where

$$\lambda_{dry} = (1-p) * p_{01} * n_m \qquad \text{Eq. 4.23}$$

Combination of Eqs. 4.19, 4.23, 4.6 $(p = p_{01}/(1-C))$ and 3.5 $(C = p_{11} - p_{01})$ results in

$$\lambda_{dry} = \lambda_{wet} = \lambda \qquad \text{Eq. 4.24}$$

Herewith the symbol λ is introduced to represent both λ_{wet} and λ_{dry}. Figs. 4.11 shows the probability density function of dry spells as a function of different transition probabilities. Figure 4.12 shows the relationship between different transition probabilities and λ.

It is theoretically possible that a month with rainfall has no dry spells, therefore the limitation of the number of dry spells is slightly different from that of wet spells:

$$0 \le N_{dry} \le \text{trunc}\left(\frac{n_m - 1}{2}\right) + 1 < \frac{n_m}{2} + 1 \qquad N_{wet} \in \mathbb{N} \qquad \text{Eq. 4.25}$$

The truncation of the Poisson distribution by the upper limit of the number of dry spells is negligible.[21]

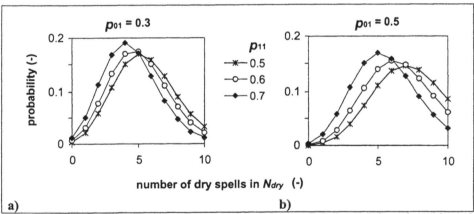

a)   b)

Figure 4.11   Probability density function of number of dry spells in a month for various values of p_{11} for a) $p_{01} = 0.3$ and b) $p_{01} = 0.5$.

---

[21] As is done for the probability density function of wet spells, the sum in the denominator is replaced by $\exp(\lambda)$ because it equals the Taylor development. This yields:

$$P\langle N_{dry} | n_m \rangle = \frac{\exp(-\lambda)(\lambda)^{N_{dry}}}{N_{dry}!} * \frac{1}{\exp(-\lambda)\exp(\lambda)} = \frac{\exp(-\lambda)(\lambda)^{N_{dry}}}{N_{dry}!}$$

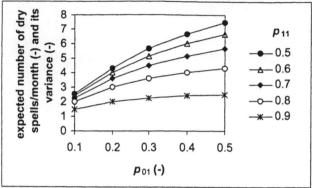

Figure 4.12   Expected number of dry spells in month (30 days) for different values of p_{01} and p_{11}. The variance of the number of dry spells/month is the same as the expected number.

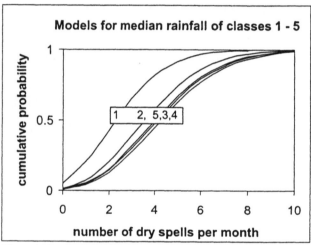

Figure 4.13   Models of cumulative distributions of dry spells or pairs of dry and wet spells for the different rainfall classes. The class with the highest rainfall (5) has fewer pairs than classes with lower rainfall.

Figure 4.13 shows that for rainfall amounts above 90 mm/month (classes 3, 4, 5), the cumulative probability distributions are very similar.

The performance of the theoretical probability density functions (Eqs. 4.18 and 4.22) is checked for Harare for dry spells - Figure 4.14 - and for wet spells - Figure 4.15. The model underestimates the number of dry and wet spells for low monthly rainfall. This can be explained by the effect of the boundaries of a month. Imagine that the expected length of a wet spell is 2 days and the expected length of a dry spell is 29 days. In a month with 31 days, the model would then expect one dry spell. However, for every historical month that the 2 rain-days are not at the end or at the start of the month, the 'measured' number of dry spells is two. Similar effects occur at wet spells. However, with low monthly rainfall the length of wet spells is shorter than that of dry spells, so the effect will be relatively small.

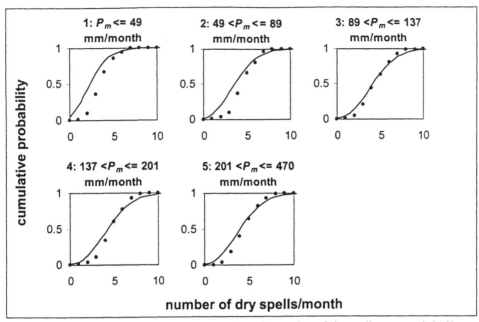

Figure 4.14   Cumulative probability distributions for number of dry spells per month in Harare (meteorological station Belvedere) for different rainfall classes (dots) and models of cumulative distributions using the Markov properties (lines). The median rainfall in the classes is used to represent the classes.

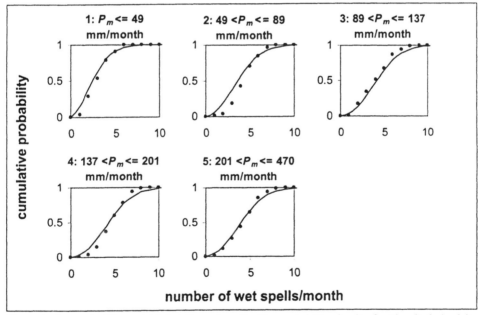

Figure 4.15   As Figure 4.14, but for wet spells. The probability density function is derived in footnote 20.

Both the mean and the median of a Poisson distribution are λ (e.g. Clarke and Cooke, 1988). Therefore:

$$\mu_{N_{dry}} = \mathrm{E}\left(N_{dry}\big|n_m\right) = \lambda$$

(-)   Eq. 4.26

$$\sigma^2_{N_{dry}} = \mathrm{Var}\left(N_{dry}\big|n_m\right) = \lambda$$

(-)   Eq. 4.27

For the number of wet spells mean and variance can be derived by deriving the first and second moment of Eq. 4.21.[22]

Figure 4.16 shows how the average number of dry and wet spells in a month develops in relation to the monthly rainfall. The theoretical expected number of wet spells per month agrees well with the observed averages for the different classes of monthly rainfall. For dry spells the performance seems worse, however this is again due to the way the number of spells in the month were determined rather than due to the model. The spells that continued beyond the end of the month, or had already started at the start of the month, were counted as 1 instead of as a portion.

By combining of Eqs. 4.25, 4.11 and 4.16 it can be shown that the number of days in a month equals the product of the expected number of dry spells and the sum of the length of an average wet spell and an average dry spell:

$$
\begin{aligned}
n_m &= \mathrm{E}(N_{dry}\big|n_m) * \left(\mathrm{E}(n_{wet}) + \mathrm{E}(n_{dry})\right) \\
&= n_m\left(1 - \frac{p_{01}}{1 - p_{11} + p_{01}}\right)p_{01} * \left(\frac{1}{p_{01}} + \frac{1}{1 - p_{11}}\right) \\
&= n_m
\end{aligned}
$$

Eq. 4.28

Thus it is concluded that the expected number of pairs of dry and wet spells is the same as the expected number of dry spells:

$$\mu_{N_{dry}} = \mu_{N_{pairs}} = \lambda$$

(-)   Eq. 4.29

where N_{pairs} is the number of pairs of a dry and wet spell in a month (1/month). It is noted that the line of reasoning above cannot be started from the expected number of wet spells, because of the fact that for wet spells the Poisson distribution is not entirely correct.

---

[22] By taking the first and the second moment of the limited Poisson distribution the mean and variance of the number of wet spells in a month are derived.

$$\mu_{N_{wet}} = \mathrm{E}\left(N_{wet}\big|n_m\right) = \frac{\lambda}{1 - \exp(-\lambda)}$$

(-)

$$
\begin{aligned}
\sigma^2_{N_{wet}} = \mathrm{Var}\left(N_{wet}\big|n_m\right) &= \frac{\lambda}{1 - \exp(-\lambda)} - \lambda^2 \exp(-\lambda) \\
&\quad - 2\lambda^2 \exp(-2\lambda) - \lambda^2 \exp(-3\lambda)
\end{aligned}
$$

(-)

Eq. 4.29 implies, in the theoretical case of the 'average month', that if a month starts with a dry spell it will finish with a wet spell and vice versa. It should be realised that the expected number of dry spells in a month is not an integer value. For example, if λ is 3.6, when the 'average month' starts with a dry spell, this will only continue for $0.6*E(n_{dry})$, because the actual spell began in the previous month. After this wet spell, three wet spells of average length $E(n_{wet})$ each followed by a dry spell of average length $E(n_{dry})$ will occur. The month finishes with a wet spell of duration $0.6*E(n_{wet})$. The expected duration of wet and dry spells is not an integer value either.

In the model for transpiration (Chapters 7 and 8) the number of pairs of a wet and dry spell in the month plays an important role. Figure 4.16 shows that the number of pairs in the month reaches a maximum at a certain monthly rainfall. With low rainfall amounts the dry spells tend to be very long, so limiting the number of pairs. With high monthly rainfall the wet spells tend to be very long, thus also limiting the number of pairs.

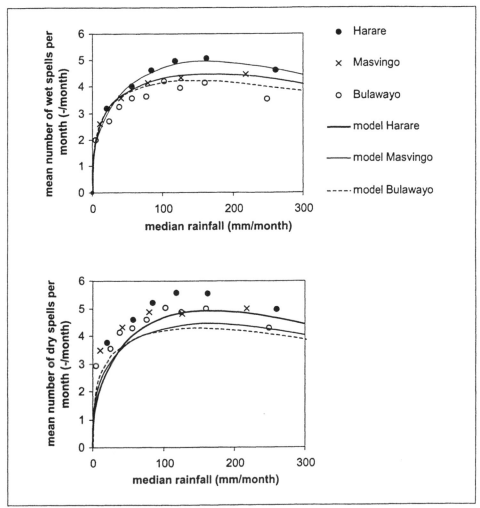

Figure 4.16   Mean number of (a) wet (b) dry spells in relation to monthly rainfall for Harare, Masvingo, Bulawayo, observed (markers) and modelled (lines).

4.6   Maximum spell length

Longest dry spell

The length of the longest dry spell affects the yield of a rainfed crop, dependent on the stage of the growing process (establishment, vegetative, heading + flowering, yield formation, De Bie, 2000). Therefore it is useful to know what the probability is that a certain length of dry spell will not be exceeded. It regularly happens that a mid-season dry spell severely diminishes crop yields (Scoones et al., 1996).

Şen (1980; see also Sharma, 1996b) determines the cumulative density function of the length of the longest dry spell $n_{dry,max}$ in a range of n_x days. Trivially, if n becomes greater, $n_{dry,\ max}$ stays the same or also becomes greater. Because the number of days n_x is discrete, the maximum length of a dry spell in n_x days is called a stochastic process with a non-decreasing step realisation, i.e. $n_{dry,max\ (n)} \le n_{dry,\ max\ (n+1)}$ for all n.

Here the theory is applied to months, thus $n_x = n_m$. The probability that the maximum length of the dry spells in a month is less than or equal to n is:

$$P\left(n_{dry,max} \le n\right) = P\left(N_{dry} = 0\right) + \sum_{i=1}^{\infty}\left[P\left(n_{dry} \le n\right)\right]^i P\left(N_{dry} = i\right)$$ Eq. 4.30

where
N_{dry} is number of dry spells   (-)
The first term, $P(N_{dry} = 0)$, is the probability that there is no dry spell at all, which implies that the maximum dry spell length is zero. The second term is a summation of all possible numbers of spells in the month. From Eq. 4.22 it follows that $P(N_{dry}=0) = \exp(-\lambda)$. Substitution of Eq. 4.22 yields:

$$P\left(n_{dry,max} \le n\right) = \exp(-\lambda)\left(1 + \sum_{i=1}^{\infty}\frac{\left[P\left(n_{dry} \le n\right) * \lambda\right]^i}{i!}\right)$$ Eq. 4.31

In reality the number of dry spells is constrained because of the limited number of days in a month, see Eq. 4.25 Large numbers of spells are very improbable and therefore

$$\sum_{i=1}^{n_m/2+1}\left[P\left(n_{dry} \le n\right)\right]^i P\left(N_{dry} = i\right) \approx \sum_{i=1}^{\infty}\left[P\left(n_{dry} \le n\right)\right]^i P\left(N_{dry} = i\right)$$ Eq. 4.32

It should be realised though that if the method is applied to time periods of less than a month ($n_x < n_m$), the inaccuracy of the approximation decreases rapidly with the number of days n to which it is applied. Fortunately, the period of a month is sufficiently long. The advantage of the endless summation is that, with some algebra, Eq. 4.31 can be written as (Şen, 1980):

$$P\left(n_{dry,max} \le n\right) = \exp\left(-\lambda * \left[1 - P(n_{dry} \le n)\right]\right)$$
$$= \exp\left(-\lambda * \left[P(n_{dry} > n)\right]\right)$$ Eq. 4.33

By substitution of Eqs. 4.10 and 4.23 in Eq. 4.33 it follows:

$$P\left(n_{dry,max} \leq n\right) = \exp\left[-n_m * \left(\frac{p_{01}}{1-C}\right) * (1-p_{11}) * (1-p_{01})^n\right]$$

Eq. 4.34

In Figure 4.17 the results for Harare Belvedere are shown. The figure shows that the model Eq. 4.34 performs poorly for the low rainfall classes. The probability of short maximum dry spell lengths is underestimated and the probability of long maximum dry spell lengths is overestimated. The underestimation of short maximum spell lengths may be due to the fact that the relationship between P_m and p_{11} was difficult to determine for low rainfall amounts. For the power function $p_{11} = u P_m^v$ it was assumed that the intercept with the $P_m = 0$ axis is 0, which is not necessarily the case. For $P \leq$ 25 mm/month the probability that maximum dry spells exceed 30 days is still 6%. For this tail end of the cumulative distribution, part of the explanation of deviations between measurements and model is that they are due to the fact that dry spells are cut off by the limits of months.

The cumulative density function of Eq. 4.34 is useful, but as a management tool it may be easier to use the maximum spell length with a certain probability of exceedance. For example, imagine that the extent of crop failure is related to the length of the maximum dry spell. In that case it is practical to express crop destruction as a function of the monthly rainfall, through the intermediate step of a relationship between maximum dry spell length, with a certain probability of exceedance, and crop failure extent.

By inverting Eq. 4.34, maximum spell lengths with a certain probability of non-exceedance F are determined:

$$n_{dry,max}\langle F\rangle = \frac{\ln\left(\dfrac{\ln(F)}{-n_m * \left(\dfrac{p_{01}}{1-C}\right) * (1-p_{11})}\right)}{\ln(1-p_{01})}$$

Eq. 4.35

To determine the median, substitute $F = 0.5$. To determine the 90% confidence limits, substitute $F = 0.95$ and $F = 0.05$. Note that $n_{dry,max}\langle 0.5\rangle$ is the spell length with a probability of non-exceedance of 0.5, which includes the case that $n_{dry,max} = n_{dry,max}\langle 0.5\rangle$. If this probability is excluded, the determined lengths immediately increase by one day. To average between the two possibilities, e.g. a probability of F that $n_{dry,max} > n$ or a probability of $(1-F)$ that $n_{dry,max} < n$, 0.5 day is added to the $n_{dry,max}\langle F\rangle$ determined in Eq. 4.35. The results are shown in Figure 4.18. At very low monthly rainfall (<10 mm/month for 50% probability of exceedance) the length of the maximum dry spell increases with monthly rainfall. This is not realistic, but is due to the fitted power functions for p_{01} and p_{11}, as in the explanation of the deviations in Figure 4.17. The most extreme maximum spell lengths are often in November or April. In the case of November, the rainy season started only late in the month. In the case of April, the rainy season ended in early April.

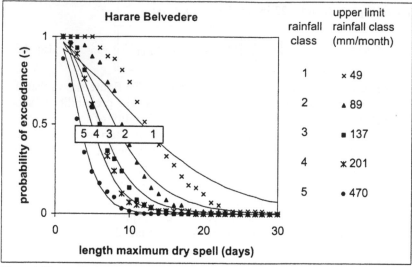

Figure 4.17   Probability of exceedance for length of maximum dry spell in month for different rainfall classes, of which the upper limit of monthly rainfall is given (e.g. class 3: $89 < P \leq 137$ mm/month). Historical values are given in dots, models in lines. For the models (Eq. 4.34) the medians of the rainfall classes are used as input.

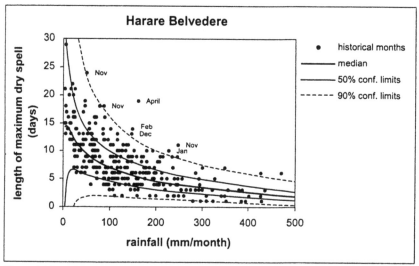

Figure 4.18   Lengths of maximum dry spells in a month for Harare Belvedere as a function of the monthly rainfall. Historical months (dots) are compared with models derived from the Markov process for certain probabilities of exceedance (lines), which form the boundaries of confidence zones.

Using Eq. 4.34, the probability that $n_{dry,max}$ is exactly n can also be derived (Şen, 1980), since:

$$P\left(n_{dry,max} = n\right) = P\left(n_{dry,max} \leq n+1\right) - P\left(n_{dry,max} \leq n\right) \qquad \text{Eq. 4.36}$$

which results in

$$P(n_{dry,max} = n) = \exp\left[-n_m * \left(\frac{p_{01}}{1-C}\right) * (1-p_{11}) * (1-p_{01})^{n+1}\right.$$
$$\left. - \exp\left[-n_m * \left(\frac{p_{01}}{1-C}\right) * (1-p_{11}) * (1-p_{01})^{n}\right]\right]$$

Eq. 4.37

Longest wet spell

The longer a wet spell is, the greater the probability will be that the soil reaches field capacity and saturation overland flow occurs. The waterlogging can destroy crops. Also, during a long wet spell sunlight often becomes the limiting resource for crop growth, instead of water. These are reasons that make the determination of the probability density function of the longest wet spell useful.

Using the knowledge $\lambda_{dry} = \lambda_{wet} = \lambda$ (Eq. 4.24), the probability that the maximum length of a wet spell is smaller than or equal to n reads:

$$P(n_{wet,max} \le n) = \exp[-\lambda * P(n_{wet} > n)]$$
$$= \exp\left[-n_m * \frac{p_{01}}{1-C} * (1-p_{11}) * (p_{11})^{n}\right]$$

Eq. 4.38

Figure 4.19 shows that for high amounts of monthly rainfall, the model overestimates the real maximum spell lengths. As in the case of the poor performance for the model of maximum dry spell lengths in months of low rainfall, the explanation lies in spells that continue beyond the limits of a month.

Of course, for the probability of exact lengths, a similar equation to Eq. 4.37 can be derived. The result is shown in Figure 4.20.

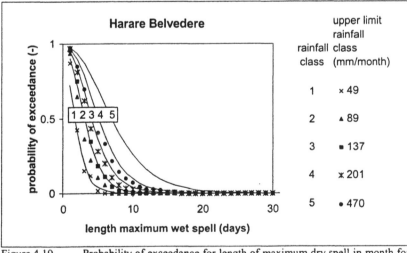

Figure 4.19   Probability of exceedance for length of maximum dry spell in month for different rainfall classes, for which the upper limit of monthly rainfall is given (e.g. class 3: $89 < P \le 137$ mm/month). Historical values are in dots, models in lines. For the models (Eq. 4.34) the medians of the rainfall classes are used as input.

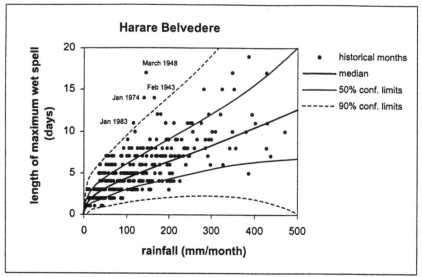

Figure 4.20   Lengths of maximum wet spells in a month for Harare Belvedere as a function of the monthly rainfall. Historical months (dots) are compared with models derived from the Markov process for certain probabilities of exceedance (lines), which form the boundaries of confidence zones.

4.7   The first rain-day

How much water a plant needs and what the effect will be of water stress depends on its cropping stage. Therefore crop yield depends heavily on the planting date. Crop yields of sorghum in the drier areas of Zimbabwe have been found to differ by a factor of 6 (3500 kg/ha instead of 600 kg/ha) depending on planting dates in the same rainy season (Nyamudeza, 1999).

Mudege (1999) used a daily soil moisture balance simulation model. For the planting date he assumed that the farmer would start planting after the first rains of the season. In models based on monthly data, such as CROPWAT (Clarke et al., 1998), the user sets the planting date as an input parameter. To determine the most probable start of the rainy season Stern & Coe (1984) used Markov stochastic models. For rainfed agriculture, the use of a fixed planting date is less realistic than the assumption made by Mudege.

The probability density function that is derived in this Section offers a simple improvement on the method used by CROPWAT. The probability that the month starts with a dry day is $(1-p)$ based on the monthly rainfall - see Eq. 4.6. The expected length of the dry spell is $E(n_{dry}) = 1/p_{01} \, \Delta t_m$ - see Eq. 4.11. The expected date of the first rain-day $n_{r,first}$, then follows through a weighted average:

$$E(n_{r,first}) = \left(p + (1 - p)\frac{1}{p_{01}} \right)\Delta t_d \qquad \text{(days)} \qquad \text{Eq. 4.39}$$

Thus if $1 < E(n_{r,first}) < 2$, the second day in the month is the expected first rain-day. Figure 4.21 shows that Eq. 4.39 is a good representation of the average date of the first rain-day in a month. The higher estimate from the averages of 32 data points is the effect of rounding off; while in the historical months rain on the second day is considered date 2, in the model these days are in the range between 1 and 2. Figure 4.21 also shows that the scatter is very large and not symmetrical around the average. This is logical, because the probability density function of dry spells is an exponential distribution. To derive planting dates, only the start of the rainy season is important. Figure 4.22 shows a scatter plot for November. The data points are too few to recognise the pattern of the model.

With the probability density function of lengths of dry spells (Eq. 4.9), for each monthly rainfall a probability density function for the date of the first rain-day can be derived, in which the skewed distribution is represented. In this way a probability density function for planting dates could be derived, which may be used to run crop yield models several times for the same season and so derive a probability density function for crop yields. As this is too cumbersome for direct practical use in monthly models, the derivation is left to the reader.

The expected crop yields are highest if crops are planted as early in the season as possible. In reality however farmers, in particular small-scale farmers, are not optimisers of expected benefits but minimisers of risk, therefore they spread sowing dates (Nyamudeza, 1999).

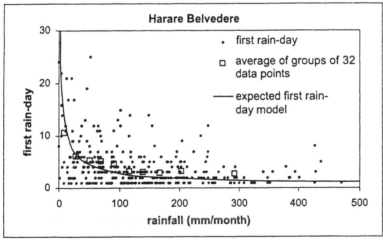

Figure 4.21   Scatter diagram of monthly rainfall against the date of the first day in the month with rainfall. The symbol + shows the averages of ten consecutive groups of 32 data points. The line shows the model derived with the Markov process for the date of the expected first rain-day.

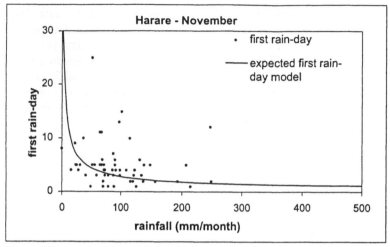

Figure 4.22   As Figure 4.21, but for November only.

4.8   Closing remarks

It has been proven that the Markov process offers a key to derive probability density functions for parameters describing the occurrence of rainfall within the month on the basis of monthly rainfall data. These parameters and examples of their use in water resources management are:

- number of rain-days   determination of interception,
- lengths of dry spells   design of supplementary irrigation systems,
- lengths of wet spells   design of rainwater harvesting systems,
- maximum length of dry spells   risk of crop destruction through water stress,
- maximum length of wet spells   risk of crop destruction through water logging
                                                         and limited energy availability (sunshine),
- first rain-day   expected planting date.

Moreover, probability density functions of the number of wet spells and dry spells in a month have been presented and it was shown that the theoretically 'average month' contains the same number of wet as dry spells. The expected number of pairs of a dry and a wet spell is useful in the derivation of the transpiration equation in Chapter 7.

The probability density functions described in this chapter follow directly from the properties of the stochastic Markov model. Traditionally, disaggregation methods are used to stochastically generate many non-unique solutions from which probability density functions are derived (Monte Carlo simulation). Such a method is too cumbersome for monthly water resources models.

5 Probabilities of Daily Rainfall Amounts

5.1 Exponential probability of exceedance

In the previous chapter the occurrence of rainfall was modelled. In this chapter the amount of rainfall on rain-days will be discussed.

The historical probabilities of rainfall exceedance in Figure 5.1a,b,c, show almost straight lines on semi-logarithmic paper, similar to the exponential model in Figure 5.2. Therefore, for every class of monthly rainfall P_m the probability of exceedance of daily rainfall P_r is in reasonable agreement with an exponential function:

$$1 - F(P_r) = \exp\left(\frac{-P_r}{\beta}\right) \qquad \text{for } P_r > 0 \text{ (mm/day)} \qquad (-) \qquad \text{Eq. 5.1}$$

where
P_r is rainfall on a rain-day. A threshold can be introduced which makes that days with only trace rainfall are not considered rain-days. But here no threshold is defined, thus $P_r > 0$.   (mm/day)
β is the scale parameter of the exponential distribution, see Box 5-A.   (mm/day)

A further discussion on the choice of this distribution and its fit to the data follows in Section 5.4. First it is shown that with an exponential probability of exceedance, simple statistical parameters for the variability of rain on rain-days can be derived.

The derivative of the probability of exceedance is the probability density function, which is also an exponential distribution

$$f(P_r) = \frac{1}{\beta}\exp\left(\frac{-P_r}{\beta}\right) \qquad (-) \qquad \text{Eq. 5.2}$$

The average rainfall on a rain-day follows as the first moment

$$\mu_{P_r} = E(P_r) = \int_0^\infty P_r \frac{1}{\beta}\exp\left(\frac{-P_r}{\beta}\right) dP_r = \beta \qquad \text{(mm/day)} \qquad \text{Eq. 5.3}$$

The median $M(P_r)$ of rainfall on a rain-day is given by $F(M)=0.5$ so that

$$0.5 = \exp\left(\frac{-M(P_r)}{\beta}\right) \qquad (-) \qquad \text{Eq. 5.4}$$

which gives

$$M(P_r) = \beta \ln(2) \qquad (-) \qquad \text{Eq. 5.5}$$

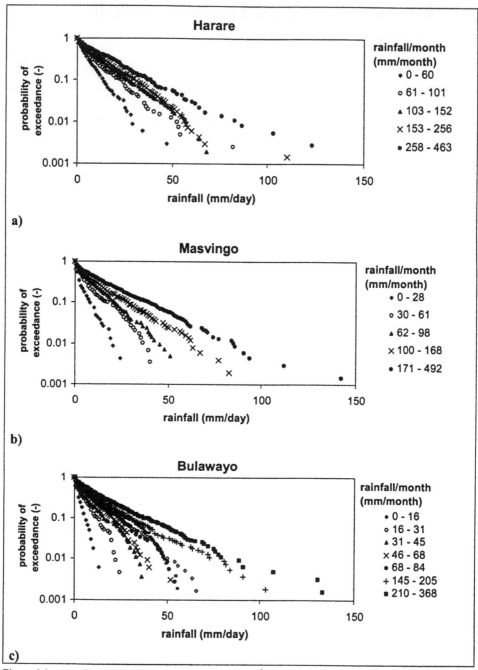

Figure 5.1   Probability of exceedance of rainfall amounts on rain-days for a) Harare, b) Masvingo, c) Bulawayo for different rainfall classes.

The variance of P_r can be expressed simply as

$$\text{Var}(P_r) = \text{E}(P_r^2) - \text{E}(P_r)^2 = 2\beta^2 - \beta^2 = \beta^2 \qquad (\text{mm/day})^2 \qquad \text{Eq. 5.6}$$

thus $C_v = \sigma/\mu = \beta/\beta = 1$. (A clear description of exponential distributions is given in Clarke and Cooke, 1988.)

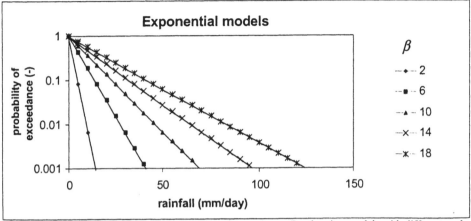

Figure 5.2   Exponential models for probability of exceedance. By showing models with different scale factors β, the sensitivity of the models to this β is illustrated. The scale parameter β is equal to the mean rainfall amount on a rain-day (mm/day).

5.2   No autocorrelation of rain on rain-days

Most stochastic models for daily rainfall generation assume that, although there is autocorrelation in rainfall occurrence, the precipitation amounts on successive days are not correlated (Woolhiser & Roldán, 1982; Wilks, 1989). Buishand (1977) studied autocorrelation of rainfall on rain-days. Amounts within a wet spell are not independent of each other. However, the correlation that he found is weak and the stations that showed correlation are in climates with monsoons (< 0.25 Calcutta, < 0.17 Jakarta). Buishand made a distinction between solitary rain-days, rain-days with another rain-day on one side and rain-days with a rain-day on both sides. Woolhiser & Roldán (1982), however, deem it unnecessary to implement dependence of rain on consecutive rain-days, as they regard the dependence found by Buishand and also by Katz (1977) as too weak. Also, Stern & Coe (1984) did not find autocorrelation in rainfall amounts on rain-days at stations from fifteen different countries.

Autocorrelation in rain on rain-days complicates the modelling approach considerably. Introducing autocorrelation in the model would introduce at least one additional parameter, which may be dependent on the location, season and monthly rainfall. For the benefit of the modelling, it is desirable that there is no autocorrelation. Here it will be explained that this is justified.

Box 5-A The simple mathematics of an exponential function

It is likely that some of the people who are interested in this dissertation have forgotten large parts of the mathematics of their secondary and undergraduate education. As the author would not like to create the impression that this dissertation contains anything difficult, some properties of the exponential equation are explained in this box. This 'refresher course' shows the simplicity (and beauty) of derivations from the exponential equation.

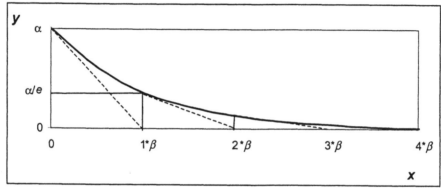

Figure 5A.1 Exponential distribution Eq.5a.1

In Fig. 5A.1 the exponential function is depicted:

$$y = \alpha \exp\left(\frac{-x}{\beta}\right)$$

Eq. 5a.1

with $\exp(x) = e^x$, and $e = 2.71828$. β is the scale parameter of the exponential function. From Eq. 5a.1 follows:

$$\int_0^\infty y \, dx = \alpha\beta$$

Eq. 5a.2

Hence the integral of the exponential function equals the size of the square enclosed by α and β. Combining Eq. 5a.1 and 5a.2, yields

$$\int_0^{x_1} y \, dx = \alpha\beta * \left(1 - \exp\left(\frac{-x}{\beta}\right)\right)$$

Eq. 5a.3

Also

$$\frac{dy}{dx} = -\frac{\alpha}{\beta} \exp\left(\frac{-x}{\beta}\right)$$

Eq. 5a.4

For $x=0$, $\exp(-0/\beta)=1$, thus $dy/dx = -\alpha/\beta$ as illustrated in Fig. 5A.1. This means that the tangent through the origin crosses the x-axis at $x=\beta$. For $x=2*\beta$ yields $dy/dx = -\alpha/\beta \exp(-2)$, which implies that the tangent at $(2\beta, \alpha \exp(-2))$ crosses the x-axis at $3*\beta$ and so on. Thus, the scale parameter β determines both the slope and the integral of the function.

To solve x from Eq.5a.1 use:

$$x = -\beta \ln\left(\frac{y}{\alpha}\right)$$

Eq. 5a.5

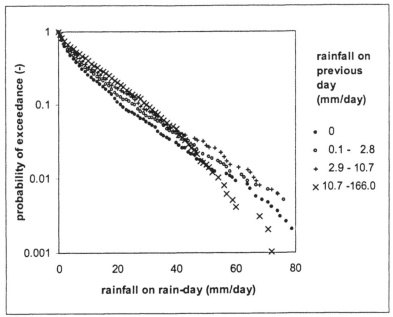

Figure 5.3   Probability of exceedance of rain on rain-days for different classes of rainfall on the previous day, using data from the meteorological station Harare-Belvedere.

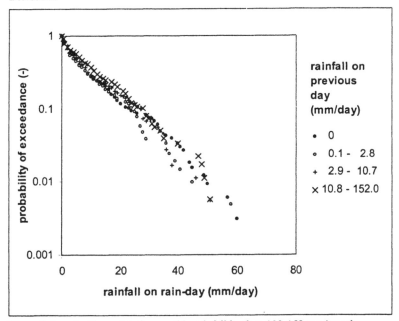

Figure 5.4   As Figure 5.3, for monthly rainfall in class 100-152 mm/month.

Figure 5.3 illustrates that there is some positive autocorrelation in amounts of rainfall on successive rain-days. When rainfall was high on the previous day, the chances are that the rainfall is high on the day itself as well (β is higher). The plot of with 0 mm/day rainfall on the previous day shows the probability of exceedance for rain on rain-days that are preceeded by dry days. Thus, comparing this line with the others illustrates the dependency on occurrence of rainfall on the previous day, in analogy with Buishand (1977). In general the correlation in rainfall amounts is positive. Only if there have been very high amounts of rainfall on the previous day does the probability of extreme rainfall on the rain-day appear to be less than if there have been low amounts of rain on the previous day.

However, autocorrelation as shown in Figure 5.3 is partly incorporated in the exponential distributions that depend on monthly rainfall. The expected rain on a rain-day increases with monthly rainfall. Thus, in months with higher rainfall, chances of high rainfall do not only apply to a particular rain-day, but also to the previous day. Figure 5.7 shows the same graph, but for months with rainfall of between 100 and 150 mm/month only. This refers to the middle class in Figure 5.1. The autocorrelation is considerably less than in Figure 5.3. No significant differences exist for the probability distributions for the different classes with rain on the previous day below 10.7 mm/day, which applies to about 70% of the rain-days.

For the above reasons, the assumption that no autocorrelation exists between rainfall amounts on rain-days seems justified and such autocorrelation is therefore ignored in all further derivations in this dissertation.

5.3   Link between occurrence of rain-days and quantities of rain on rain-days

In this Section it is shown that the scale factor β can be derived from the parameters of the Markov process, which describe the rainfall occurrence.

Expressing β as a function of the Markov process parameters

In the previous Section it was derived that the scale factor β is equivalent to the mean rainfall on a rain-day (Eq. 5.3)

$$\beta = \mu_{P_r} \hspace{4cm} \text{(mm/day)} \hspace{1cm} \text{Eq. 5.7}$$

Trivially, the monthly average rainfall on rain-days equals the ratio of the monthly rainfall to the number of rain-days in the month;

$$\overline{P_r} = P_m / n_r \hspace{4cm} \text{(mm/day)} \hspace{1cm} \text{Eq. 5.8}$$

The expected mean rainfall on a rain-day is the same as the expected average rainfall on a rain-day:

$$\mu_{P_r} = E(\overline{P_r}) \hspace{4cm} \text{(mm/day)} \hspace{1cm} \text{Eq. 5.9}$$

Drawing on the properties of the Markov process, in Section 4.2 a probability density function for the number of rain-days in a month (n_r) is described as a normal distribution, with mean as in Eq. 4.2 and variance as in Eq. 4.3. Thus, the probability density function for β can be described as a function of monthly rainfall, by taking $\beta = P_m/n_r$ and by expressing the probability density function of n_r as a function of P_m. The uncertainty in the number of rain-days per month causes an uncertainty in the average rainfall on rain-days. The monthly rainfall P_m is known and thus does not have any variance, but the number of rain-days n_r is not.

Thus have been described all ingredients for an unconditional stochastic downscaling model. The monthly rainfall supplies the model parameters of the stochastic model for daily time steps, but does not put conditions to the monthly rain-sum of the daily time steps.[23] In this dissertation disaggregation of the monthly totals to daily time series is not used. Instead, the statistical parameters of the daily variability are used as model parameters in the monthly equation. However, by applying the model characteristics in an unconditional stochastic model, for each starting value of monthly rainfall P_m histograms of generated monthly rainfall can be created, which are illustrative for the model characteristics. This is shown in Figure 5.5.

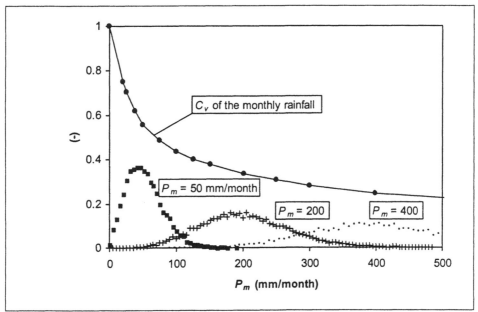

Figure 5.5   Histograms of the monthly rain totals of stochastically generated time series using the assumptions of this dissertation, for P_m = 50, P_m = 200, P_m = 400 mm/month and modelled transition probabilities from Harare (for P_m = 50, p_{01} = 0.17, p_{11} = 0.51, β = 6.4 mm/day; for P_m = 200, p_{01} = 0.37, p_{11} = 0.71, β = 11.9 mm/day; for P_m = 400, p_{01} = 0.54, p_{11} = 0.84, β = 17.2 mm/day). Also the coefficient of variation is shown as a function of the monthly rainfall: $C_v = \sigma/\mu$.

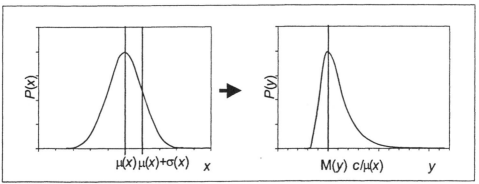

Figure 5.6   a) Probability density function of x, which is a normal distribution, b) Probability density function of y, where $y=c/x$ and c is a constant.

---

[23] See for a lucid explanation on scale transfers Bierkens et al. (2000). Instead of using conditional stochastic models, some other researchers have used unconditional stochastic models to disaggregate monthly rainfall. Thornton et al. (1997) multiplied the daily rainfall amounts on rain-days with a monthly correction factor. Mulligan & Reaney (1999) generated many months of daily time series until a monthly rain sum was similar to the monthly value to be disaggregated.

The difference between median and mean rainfall amount on rain-days

For any number of rain-days which is greater than $E(n_r)$, the average rainfall amount on rain-days is less than $P_m/E(n_r)$ and vice versa. Thus, the reciprocal value of a variable with a normal distribution is a skewed variable, with a median that is the reciprocal value of the mean of the variable of the normal distribution. See Figure 5.6. Therefore, for a given monthly rainfall the median of average rainfall on a rain-day reads:

$$M(\overline{P_r}) = \frac{P_m}{E(n_r|n_m)} = P_m * \frac{(1-C)}{n_m P_{01}} \qquad \text{(mm/day)} \qquad \text{Eq. 5.10}$$

The derivation of the mean is more complex. For a quotient c/x, where c is a constant ($= P_m$) and x agrees to a normal distribution ($= n_r$), Taylor series expansion about (c, μ_x), neglecting terms higher than order 2, yields the following estimate (Mood, Graybill & Boes, 1963):[24]

$$E\left(\frac{c}{x}\right) \approx \frac{c}{\mu_x} + \frac{c}{(\mu_x)^3} \text{Var}(x) \qquad \text{(mm/day)} \qquad \text{Eq. 5.11}$$

The problem of determining the mean of $c/x = P_m/n_r$ offers some constraints:
- $0 < n_r < 31$, thus x is limited,
- $0 < P_m$, thus $c > 0$,
- for Harare, Masvingo, Bulawayo $\sigma_{n_r} < 6$.

Substitution of $c = P_m$ and $\mu_x = E(n_r)$ (Eq. 4.4) and $\text{Var}(x) = \text{Var}(n_r)$ (Eq. 4.5) yields:

$$E(\overline{P_r}) = E\left(\frac{P_m}{n_r}\right) \approx \frac{P_m}{n_m} \frac{1-C}{P_{01}}$$

$$+ P_m \left(\frac{1}{n_m P_{01}}\right)^2 (1-C)(1+C)\left(1 - \frac{P_{01}}{1-C}\right) \qquad \text{(mm/day)} \qquad \text{Eq. 5.12}$$

$$= M\left(\frac{P_m}{n_r}\right) * \left\{1 + \frac{1}{n_m P_{01}}(1+C)\left(1 - \frac{P_{01}}{1-C}\right)\right\}$$

The reciprocal transformation of the variable with a normal distribution n_r shows that:

$$E(\overline{P_r}) > M(\overline{P_r}) \qquad \text{(mm/day)} \qquad \text{Eq. 5.13}$$

This is by definition true, because the factor between {} in Eq. 5.12 is always more than 1. Eq. 5.12 is an estimate of the mean. In particular for small monthly rainfall amounts, where the standard deviation in the number of rain-days is large in comparison to the mean number of rain-days (see Figure 4.2 a,b,c), underestimation of $E(\overline{P_r})$ can be as great as 25%.

Figure 5.7 presents the ratio of the mean to the median of rainfall on a rain-day for Harare, Masvingo and Bulawayo respectively. For very low monthly rainfall the relative difference tends towards infinity.

---

[24] An estimate of the variance is:

$$\text{Var}\left(\frac{c}{x}\right) \approx \left(\frac{c}{\mu_x}\right)^2 * \frac{\text{Var}(x)}{(\mu_x)^2} \qquad \text{(mm/day)}$$

but if $c/x = P_m/n_r$ the inaccuracy with small P_m is very great (Mood, Graybill & Boes, 1963).

A good estimation of the monthly mean rainfall on a rain-day will prove useful when it is necessary to derive the mean instead of the median interception. Although Eq. 5.12 is not exact and far more complex than Eq. 5.10, it is in any case a better estimation of $E(\overline{P_r})$ than can be achieved using the median.

In Figure 5.8 the drawn line represents the relationship between monthly rainfall and $M(\overline{P_r})$ and the broken line that between monthly rainfall and $E(\overline{P_r})$. The figure shows that the difference between $M(\overline{P_r})$ and $E(\overline{P_r})$ depends on the monthly rainfall and on the location, but is less than 15% for monthly rainfall greater than 100 mm/month.

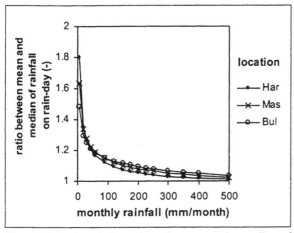

Figure 5.7   Ratio of the estimated mean to the median of rainfall on rain-day as a function of monthly rainfall for Harare, Masvingo and Bulawayo.

The confidence limits of the average rainfall amount on rain-days

Through numerical integration it is possible to derive confidence limits for the monthly average of rainfall on rain-days. In Figure 5.8a,b,c the 90% confidence limits are shown. Both the historical values and the theoretically-derived confidence limits illustrate the skewness of the probability density functions of $\overline{P_r}$, for a given P_m. For small values of monthly rainfall, the probability that there is only 1 rain-day is larger than 5%, therefore the higher confidence limit of 90% is in that case the total monthly rainfall in one day.

The modelled confidence limit of 90%, which has a 95% probability of exceedance, is convincingly confirmed by the historical measurements. For the 5% probability of exceedance limit this is more difficult to judge. Figure 5.8 shows that it is important to be aware of the skewness in average rain on rain-days. Average rain on rain-days can be far more than the median and the mean of the expected rainfall. In comparison to the large difference between the confidence limits, the difference between the median (Eq. 5.10) and the mean (Eq. 5.12) is small. The fact that a month with rainfall has at least one rain-day constrains the confidence limit with low probability of exceedance for small rainfall amounts. This limit is not intersected by the median (see Figure 5.7), but is intersected by the mean. This is an additional argument for using the median.

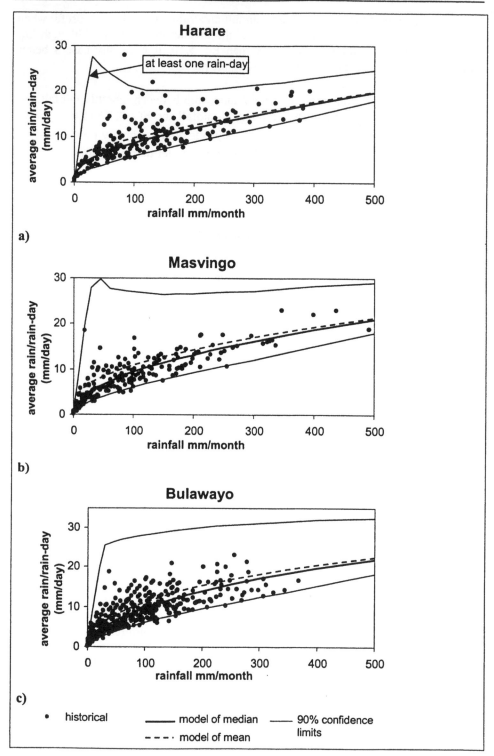

a)

b)

c)

Figure 5.8 a,b,c   Monthly average of rainfall per rain-day for historical months and for model of median and mean based on Markov process, for a) Harare, b) Masvingo, c) Bulawayo. The thin lines give the modelled 90% confidence limits.

Concluding remarks

For the derivation of β the median of the average rainfall on rain-days is used, for reasons of simplicity and transparency. Combining Eq. 5.3 and Eq. 5.10, the scale factor β can be expressed as a function of the monthly rainfall P_m:

$$\beta = \mathrm{M}(\overline{P_r}|P_m) = \frac{P_m}{\mathrm{E}(n_r|n_m)} = \frac{P_m(1-C)}{n_m p_{01}} \qquad \text{(mm/day)} \qquad \text{Eq. 5.14}$$

This permits an easy determination of β as a function of P_m on the basis of the power functions between the monthly rainfall and the transition probabilities p_{11} and p_{01}.

As was shown in Section 3.7, the parameters that connect transition probabilities and monthly rainfall (q, r, u, v) do vary greatly in space. Therefore it is possible to regionalise these parameters, which makes having time series of daily rainfall measurements at all rain gauges redundant. If a better estimate of probability density function of rainfall on rain-days should be necessary, merely recording days as dry or rain-days is in most cases sufficient.

A person can easily remember which were the dry and which the rain-days in the past week. Thus, to keep record of occurrence of rain will only require a few minutes per week from him/her and does not need equipment. In remote areas, this is a far easier method than using a raingauge, which has to be properly installed and read and emptied daily. Study of some daily records of rain gauges in communal lands, showed that around Christmas there was remarkably little rainfall. A record merely of the occurrence of rain will be more reliable. In this section and in Chapter 3 it is shown that it offers very valid information.

5.4 Choice of probability density function

In Chapter 2 it has been discussed that stochastic models for rainfall generation use gamma, Weibull or exponential equations for the probability of exceedance of daily rainfall on rain-days (see Appendix B). In Section 5.1 it has been shown that common statistical properties can be derived easily from the exponential distribution. An advantage of the gamma and the Weibull equations, however, is that they give a higher probability for lower rainfall amounts per day than the exponential distribution, which generally fits better to historical data. As referred to earlier, Figure 5.1a,b,c shows the historical probabilities of exceedance of rain on rain-days for Harare, Masvingo and Bulawayo respectively. In Figure 5.1a,b,c straight lines can generally be drawn through points corresponding to probabilities of exceedance of, say, between 0.05 and 0.8. These lines in many cases cross the 0 mm/day-axis at a point below 1. This shows that in Zimbabwe too a skewed distribution would fit the data better.

The Weibull and the gamma distributions, however, have disadvantages in comparison to an exponential distribution:
a) it is more difficult to derive the moments of the probability density function,
b) more parameters are involved, which cannot be derived directly from moments of the probability density function.
In the following paragraphs these aspects are discussed further.

To derive analytical relations for water resources models with a monthly time step, moments of the probability density function of rain on rain-days need to be derived. For example, in the next chapter it will be shown that to derive monthly interception it is necessary to solve the first moment of the probability density function, namely $\int P_r{}^*$ $f(P_r)\,dP_r$. This is far easier with exponential distributions.

As shown in the previous Section, the property of the exponential distribution that the scale parameter of the exponential distribution β is the mean rainfall (first moment), makes it possible to couple the Markov process to the probability of exceedance of rain on rain-days. For the Weibull distribution it is not possible to express the parameters as functions of moments (Benjamin & Cornell, 1970), although it is possible to do so the other way around (Appendix B). Therefore, when using the Weibull distribution, it is not possible to express the probability of exceedance of rain on rain-days as a function of the Markov transition probabilities.

A good alternative to overcome the disadvantage of the exponential equation and to obtain more skewness, is to take the sum of two exponential equations, which is called the mixed exponential distribution:

$$1 - F(P_r) = \psi \exp\left(\frac{-P_r}{\beta_1}\right) + (1-\psi)\exp\left(\frac{-P_r}{\beta_2}\right) \qquad \text{Eq. 5.15}$$

where ψ is the point at which on semi-log paper the straight line that fits the points with probabilities of exceedance of between 0.05 and 0.8 would cross the $P_m = 0$-axis.

A number of authors have used the mixed exponential distribution for stochastic models of daily rainfall (Smith & Schreiber, 1974; Woolhiser & Pegram, 1979; Foufoula-Georgiou & Lettenmaier, 1987; Woolhiser et al., 1993). All models were applied in the USA, although in different climatic regions. Woolhiser & Roldán (1982) compared the mixed exponential distribution with the exponential distribution and the gamma distribution for five rainfall stations in different climates in the US. The mixed exponential distribution proved superior. Woolhiser et al. (1993) correlated the parameters of the mixed exponential distribution (and the Markov chain transition probability p_{01}) to the index of the Southern Oscillation, which is a large-scale interannual variation of barometric pressures that influences the precipitation in the tropics and subtropics.

The first moment of the mixed exponential distribution can be derived as easily as the first moment of the exponential distribution. The average rainfall on a rain-day will be similar to the exponential distribution (Eq. 5.3), namely:

$$E(P_r) = \psi\beta_1 + (1-\psi)\beta_2 \qquad \text{Eq. 5.16}$$

The variance of rainfall on rain-days is more complex. Foufoula-Georgiou & Lettenmaier (1987) present:

$$\text{Var}(P_r) = \frac{\psi}{\beta_1} + \frac{1-\psi}{\beta_1{}^2} + \psi(1-\psi)\left(\frac{1}{\beta_1} - \frac{1}{\beta_2}\right)^2 \qquad \text{Eq. 5.17}$$

The median cannot be expressed analytically. However, variance and median are not required for the derivation of interception and transpiration.

With regard to point b) mentioned above, Weibull, gamma and the mixed exponential distribution use more than one parameter (see Appendix B). As is shown in the previous Section, the scale factor β of the exponential function is the mean rainfall on a rain-day. This mean rainfall on a rain-day can be derived directly from the Markov process, which is not the case with the parameters of the other distributions. The disadvantage of using more than one parameter weighs heavily when spatial interpolation has to be done, which is the case here. The mixed exponential distribution becomes an option if β_1, β_2 and ψ can be linked, for example by fixed ratios, or if a model is set up for a small area for which a historical series of daily rainfall in the neighbourhood is representative. But still, the derivations will be less transparent. In this dissertation the exponential distribution is used, although also for the mixed exponential distribution the monthly interception equation is derived (Chapter 6).

5.5   Comparison of fitted and Markov-derived β

The previous Section explained why for practical reasons a simple exponential probability of exceedance is preferred. In this Section it is investigated whether this distribution is sufficiently accurate for the reproduction of mean, median and standard deviation of rain on rain-days. Mean, median and standard deviation of an exponential function of a β that is derived with the Markov process (Section 5.3) and of an exponential function derived with a fitted β are compared with mean, median and standard deviation in historical rainfall on rain-days for the different classes of monthly rainfall. However, first it is defined how the scale parameter β was fitted.

Lower rainfall amounts generally have a higher probability than the exponential distribution suggests. For this reason a Weibull, gamma or mixed exponential distribution is often applied. For the same reason, estimates of β by graphical fitting can easily differ 0.5 mm/day, dependent on which part of the range in daily rainfall values is emphasised. To avoid subjective fitting, but not to exaggerate the influence of the points of low rainfall, it was chosen to maximise the value for the Probability Plot Correlation Coefficient r, a parameter used in flood frequency forecasting (developed by Filliben, 1975, see Stedinger et al., 1997). If there was more than one rain-day in a class of $\Delta 1$ mm/day – e.g. the class of all rain-days with more >0 and <= 1 mm/day – this was counted as one rain-day. The reason is that for low amounts of daily rainfall there were that many rain-days that they determined the fit too much.

Figure 5.9 presents the relative errors after the fit. All rainfall classes for all three locations show the same pattern: for small daily rainfall amounts the model overestimates the probability of exceedance, for large daily rainfall amounts, the model underestimates the probability of exceedance. Although for high daily rainfall amounts the relative difference between model and measured is more severe, this is more then compensated by the fact that confidence band in the estimate is also larger for higher daily rainfall amounts. For low monthly rainfall the estimates of the measured probability of exceedance are already very uncertain at low daily rainfall amounts, as the 95% confidence band is very wide.

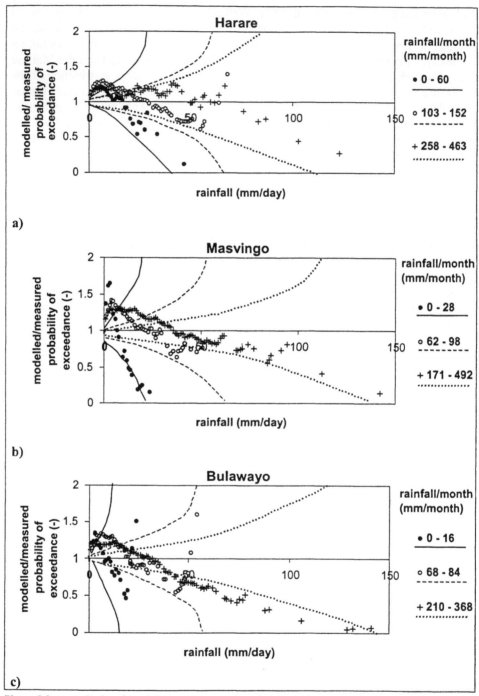

Figure 5.9   Ratio of modelled to measured probability of exceedance for different monthly rainfall groups for a) Harare b) Masvingo c) Bulawayo. The lines show the modelled 95% confidence limits for the different groups, based on a binomial distribution of the exceedance at a certain rainfall amount per rain-day. The widest limits (left solid lines) are for the group with the lowest monthly rainfall (0-60 mm/month).[25]

As expected, the probability of low rainfall amounts is considerably underestimated, while most rain-days occur in this area. In Section 5.4 the preference for an exponential function is explained. Despite the poor performance at low rainfall amounts, it is elected to adhere to the choice of an exponential distribution and to compare the mean, median and variance with observed mean, median and variance.

Comparison of mean, median and variance

With the Markov chain the mean number of rain-days per month can be determined for the median of a class of monthly rainfall. By curve fitting, β can be determined for a class of monthly rainfall. If the product of the two is a good estimate of the median monthly rainfall in the class, then the Markov β (see Eq. 5.14), is a good alternative to the fitted β. Figure 5.10 shows that good estimates of the measured monthly rainfall are indeed obtained.

Figure 5.11a,b,c shows that βs derived by the Markov process are not necessarily less appropriate than βs derived by fitting to represent the statistical parameters for rain on rain-days. For the mean rainfall on a rain-day, the Markov derived βs perform better. The fitted βs overestimate the average rainfall on rain-days by some 20%. Both types of βs overestimate the median rain on rain-days. This is logical, because the compromise was made to neglect the comparatively high frequency of low rainfall amounts. Still, substitution of the Markov βs in Eq. 5.4, will give a better estimate of the median than substitution of the fitted βs. For the standard deviations of rain on rain-days, it is the other way round. Fitted βs perform better than Markov βs. Markov βs underestimate the standard deviation by about 30%. While realising this, the use of Markov-derived βs is acceptable and preferred.

---

[25] In the figure Weibull's plotting positions are used. This is an unbiased plotting position for all kinds of distributions and equals the mean of a beta distribution (Stedinger et al., 1997):

$$E(F(P_r)) = \frac{i}{n+1}$$

for the rain-day with the ith largest total of rain and for n rain-days in the group of months considered. According to a beta distribution, the variance of the error in the probability of exceedance $F(P_r)$ is

$$\text{Var}(F(P_r)) = \frac{i(n-i+1)}{(n+1)^2(n+2)}$$

This implies that for a small probability of exceedance you would need a larger sample to get the same relative accuracy although there are less data available. In the figure the 95% confidence limits are shown, thus two times the standard relative error is plotted.

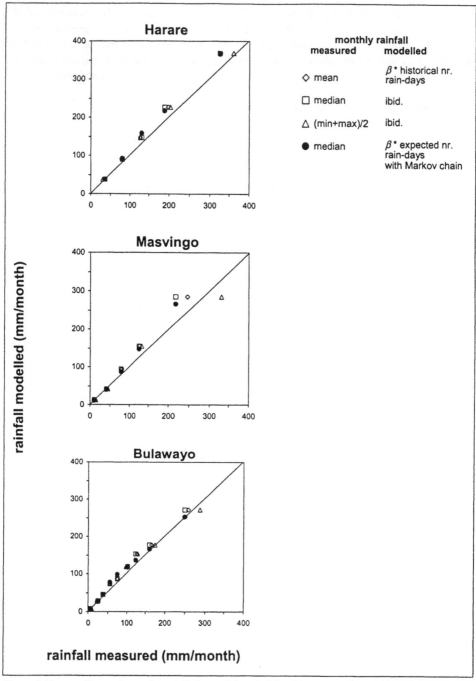

Figure 5.10   Comparison of modelled monthly rainfall with measured monthly rainfall. Model: multiplication of expected average rain on rain-day (β), derived through fit of exponential distribution through probability of exceedance, with expected number of rain-days. The number of rain-days is either counted from historical series or derived by Markov chain for the mean monthly rainfall (see Chapter 3). For measured monthly rainfall of each class mean, median or (min+max)/2 are used. a) Harare b) Masvingo c) Bulawayo.

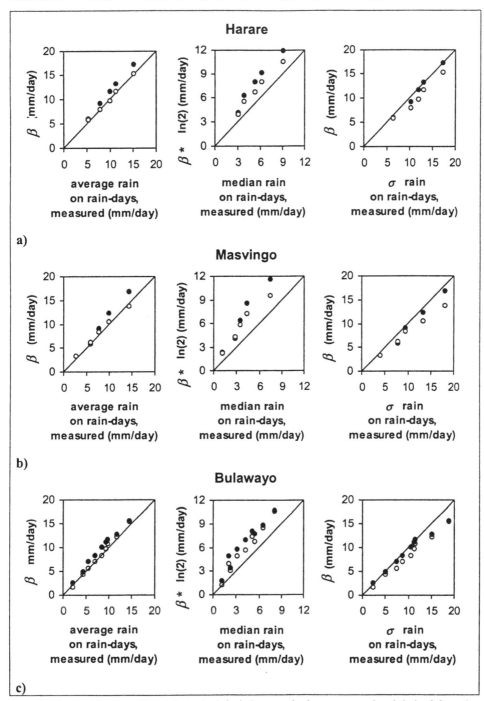

Figure 5.11   Mean, median and standard deviation on rain-days, measured and derived from the exponential model for βs, deduced by fitting (•) or by the Markov process (o). Each point represents a monthly rainfall class, with classes of higher rainfall having higher βs. a) Harare b) Masvingo c) Bulawayo.

5.6   Variability of rain when including all days

The above analysis yields daily rainfall statistics for rain-days. Sharma (1996a) gives the links between probabilities on rain-days (subscript r) and those valid for all days (subscript d), which presumes that a dry day is really dry (0 mm/day, see box 2-A). After Sharma (1996a), but quite obviously, it implies:

$$E(P_d) = E(P_r) * p$$ Eq. 5.18

where
P_d is rain on any day, including dry days   (mm/day)
p the probability of a rain-day, as derived in Eq. 4.4, $p = p_{01}/(1-C)$.

The variance of the amount of rain on all days, including dry days, is (Sharma, 1996a)

$$\text{Var}(P_d) = \frac{p * \text{Var}(P_r) + (1-p) * E^2(P_r)}{p^2}$$ $(\text{mm/day})^2$ Eq. 5.19

This is independent of the density distribution of rain on rain-days (e.g. gamma, Weibull, exponential).

A probability density distribution for rain on rain-days that is exponential, implies:

$$\boxed{\text{Var}(P_d) = \frac{\beta^2}{p^2} = \frac{P_m^2}{n_m^2 p^4}}$$ $(\text{mm/day})^2$ Eq. 5.20

5.7   Rain totals of wet spells

Knowing the statistics of totals of rainfall during wet spells will prove useful when deriving the relation for monthly transpiration. Also, probability density functions for rainsums during wet spells are useful in determining the risk of waterlogging and in determining the recuperation of the soil moisture content during wet spells, and in designing of rainwater harvesting systems.

As mentioned, Şen (1977) derived methods for annual flow series, which can be applied to rainfall series. For static, two state first-order Markov processes and exponential probability density functions for rain on rain-days, the total rain during a wet spell P_{wet} yields:

$$\boxed{E(P_{wet}) = E(P_r) * E(n_{wet}) = \frac{\beta}{1 - p_{11}} \Delta t_d}$$ (mm)   Eq. 5.21

The length of a wet spell follows an exponential distribution, because Eq. 4.14 can be transformed into:

$$P(n_{wet} > n) = p_{11}{}^n = \exp(n * \ln(p_{11}))$$ Eq. 5.22

Because rainfall amounts on rain-days have no autocorrelation, the cumulative density function for total rain during wet spells is also exponential. The scale parameter is equal to the mean rainfall in a wet spell, following Eq. 5.22:

$$1 - F(P_{wet}) = \exp\left(\frac{-P_{wet}(1 - p_{11})}{\beta \, \Delta t_d}\right)$$ (-)   Eq. 5.23

$$f(P_w) = \frac{(1 - p_{11})}{\beta \, \Delta t_d} \exp\left(\frac{-P_{wet}(1 - p_{11})}{\beta \, \Delta t_d} \right) \qquad (\text{-}) \qquad \text{Eq. 5.24}$$

Although the probability density function for lengths of wet spells is discrete, for the rain totals it is continuous, because rain on rain-days follows a continuous distribution.

Sharma (1996b) used probability distributions of total rain in wet spells for rainwater harvesting designs in Kenya. However, he used a Weibull distribution (see Appendix B).

5.8   Conclusions

In this chapter the following conclusions have been drawn:

- For a certain monthly rainfall, the probability density function of rainfall amounts on rain-days is an exponential function.

- For a certain monthly rainfall, autocorrelation of rainfall amounts on rain-days is negligible.

- The scale parameter of the exponential function β is the mean rainfall amount on a rain-day and can therefore be expressed as a function of the Markov transition probabilities and the monthly rainfall.

- The properties of the exponential function mean that the scale parameter β is also the standard deviation in rainfall amounts on rain-days and $\beta * \ln(2)$ is the median rainfall amount on rain-days.

- The βs derived by the Markov transition probabilities are performing at least as well in estimating average rain on rain-days as those derived by fitting the exponential function to the cumulative probability density function of a class of monthly rainfall. The average rainfall on rain-days and the median rainfall on rain-days are underestimated more severely by the fitted β than by the Markov-derived β. Variance in rain on rain-days is overestimated more by the Markov derived β than by the fitted β.

- The high probability of occurrence of small rainfall amounts is better represented by a mixed exponential distribution. However, this introduces two additional parameters that have to be calibrated. Only if all three parameters can be linked, for example by fixed ratios, can the mixed exponential distribution be expressed as a function of monthly rainfall.

- The expectation and variance of rainfall amounts on any day, including dry days, and the probability density function of total rain in wet spells can be directly derived from the Markov transition probabilities.

6 Interception

6.1 Introduction

The monthly interception equation that is derived in this chapter is based on the assumptions belonging to a daily model, with which it will be compared.

In this dissertation interception is defined as the amount of rain that evaporates on the day it falls. Therefore at a daily time step it can be considered as an evaporation flux. This is different to the definition of interception in the well established, physically-based Rutter model (Rutter et al., 1971). In the Rutter model the interception is the stock of water which the vegetation canopy or other surfaces carry. If interception can be considered as an evaporation flux on a daily time scale, it can certainly be considered as a flux on a monthly scale. Thus, while strictly speaking the term evaporation from interception should be employed, for notational convenience just interception is used.

6.2 Daily threshold

In this subsection the assumptions of the daily model are explained.

For the rainfall runoff process, interception is defined as that part of the rainfall that does not enter into any of the stocks contributing to transpiration or to the runoff process (S_{soil}, $S_{groundwater}$, $S_{surface\ water}$). However, the approach introduced in this chapter does not exclude a more confined method which only accounts for interception from canopy. The only condition for the daily potential interception is that the threshold D is fixed within a month. Thus, the threshold D could be regarded as either a) the mean interception capacity of the canopy only or b) the mean daily interception capacity over the surface area. The values of D may vary over the season or depend on meteorological conditions. This will not affect the approach in the next subsections.

For canopy interception only, the equation reads:

$$D = \mathrm{Min}(D_v, I_{pot,r})$$

Eq. 6.1

D_v interception capacity of vegetation   mm/day
$I_{pot,r}$ potential evaporation from wet vegetation on a rain-day   mm/day

The capacity of the canopy to intercept water varies from 0.5 mm/day for grass to some 2 mm/day for coniferous forest. However, several rainstorms or prolonged rainstorms during the day may cause more water to be evaporated from the leaf area per day (Schellekens et al., 1999).

For the average interception over the surface area, the weighted average of soil evaporation and canopy interception should be taken, disregarding storage in pools.

$$D = a \operatorname{Min}(D_v, I_{pot,r}) + (1-a) * E_{pot,s}$$ Eq. 6.2

where

$E_{pot,s}$	is potential soil evaporation	(mm/day)
a	fraction of land surface covered by vegetation	(-)

As mentioned above, it has been implicitly assumed that soil evaporation and interception from the canopy have been completed within a day after the storm. Soil evaporation occurs mainly from the top few centimetres of the soil. When these have dried out, the hydraulic conductivity of the top soil reduces considerably, which hampers the capillary rise of water to the surface. Obviously, physical parameters of the soil play a deciding role.

Pitman (1973) also assumed that the interception storage was completely evaporated between successive rain-days and that this did not affect potential transpiration. For South Africa, he considered a range of D of 0 to 8 mm/day, which is also considered suitable here. However, his calibration of streamflow output of twenty-three catchments in South Africa revealed that $D = 1.5$ mm/day could be used for the whole country, except for one densely vegetated area for which 8 mm/day was calibrated. The current established estimates for South Africa are for canopy interception thresholds of 1 to 2 mm/day. For interception thresholds that include litter interception, current estimates are up to a maximum of about 7 mm/day (Prof. Schulze, University of Pietermaritzburg, developer of the daily ACRU model, personal communication, 2001). Smit & Rethman (2000) measured plots (65 x 180m) with different densities of *Hardwicka mopane* trees, common indigenous vegetation in arid savanna in Zimbabwe, the North Province of South Africa, Mozambique and Namibia. For tree densities between 10 to 100% they concluded from soil water measurements that 40 to 60% of the rain turns into initial losses (interception + runoff). They claim that most of these initial losses are probably runoff. However, as average runoff coefficients in the region are never larger than 10%, this author suspects that interception is a major factor.

It is noted that a suitable deterministic method to determine the daily threshold D from landcover maps is difficult. Savenije (1997) developed a multiple linear regression model that expresses monthly runoff as a function of the effective rainfall in previous months. In Savenije's model monthly interception is assumed to amount the minimum of monthly rainfall and a calibrated constant monthly maximum. The daily threshold D, which is input parameter in the interception model of this dissertation, can be assessed through adapting Savenije's model. The monthly maximum in Savenije's model is replaced by the interception equation of this dissertation. In that way, the daily threshold D instead of the monthly maximum interception is calibrated. This D is a representative threshold for the whole catchment upstream of the runoff gauge that is used. It is realised that this does not necessarily give the areal mean of the daily threshold D, because of non-linearity. However, the advantage of such an approach is that only monthly rainfall and runoff data are necessary.

The determination of the potential interception will be discussed in Chapter 9. The most established and advanced model for interception is the Rutter model. Zeng et al. (2000) used this model to approximate monthly interception. A comparison with the approach presented here is made in subsection 6.7.

6.3 Analytical equation for monthly interception

Assuming that interception is the amount of rainfall that does not pass a daily threshold D, the above deduced equations can be used to derive an equation between monthly rainfall and monthly interception.

$$I_m = E(n_r|n_m) * \overline{I_r} \qquad \text{(mm/month)} \qquad \text{Eq. 6.3}$$

where I_m is the monthly interception and $\overline{I_r}$ is the average interception on a rain-day.

The rainfall on average caught in the threshold on a rain-day is by definition:

$$\overline{I_r} = \int_0^D P_r * f(P_r)\,dP_r + \int_D^\infty D * f(P_r)\,dP_r \qquad \text{(mm/day)} \qquad \text{Eq. 6.4}$$

where the first term represents rain-days for which rainfall is less than the threshold D and the second term represents rain-days for which rainfall is more than the threshold. As above, $f(P_d)$ is again the density function of the daily rainfall, which is a function of the monthly rainfall. Eq. 6.4 can be rewritten as

$$\overline{I_r} = \int_0^D \frac{P_r}{\beta}\exp\left(\frac{-P_r}{\beta}\right)dP_r + D * (1 - F(D)) \qquad \text{(mm/day)} \qquad \text{Eq. 6.5}$$

where

$$\int_0^D \frac{P_r}{\beta}\exp\left(\frac{-P_r}{\beta}\right)dP_d = \left[-P_r\exp\left(\frac{-P_r}{\beta}\right)\right]_0^D - \int_0^D -\exp\left(\frac{-P_r}{\beta}\right)dP_r$$

$$= \left[-P_r\exp\left(\frac{-P_r}{\beta}\right)\right]_0^D - \left[\beta\exp\left(\frac{-P_r}{\beta}\right)\right]_0^D \qquad \text{Eq. 6.6}$$

$$= (-\beta - D)\exp\left(\frac{-D}{\beta}\right) + \beta$$

A similar equation can be derived when using the mixed exponential distribution for the probability of exceedance.[26]

Furthermore, as a consequence of Eq. 5.1:

$$D * (1 - F(D)) = D\exp\left(\frac{-D}{\beta}\right) \qquad \text{Eq. 6.7}$$

---

[26] When using a mixed exponential distribution, the equation is:

$$\int_0^D P_r\left\{\frac{\psi}{\beta_1}\exp\left(\frac{-P_r}{\beta_1}\right) + \frac{1-\psi}{\beta_2}\exp\left(\frac{-P_r}{\beta_2}\right)\right\}dP_r$$

$$= \left[\psi\{-P_r - \beta_1\}\exp\left(\frac{-P_r}{\beta_1}\right) + (1-\psi)\{-P_r - \beta_2\}\exp\left(\frac{-P_r}{\beta_2}\right)\right]_0^D$$

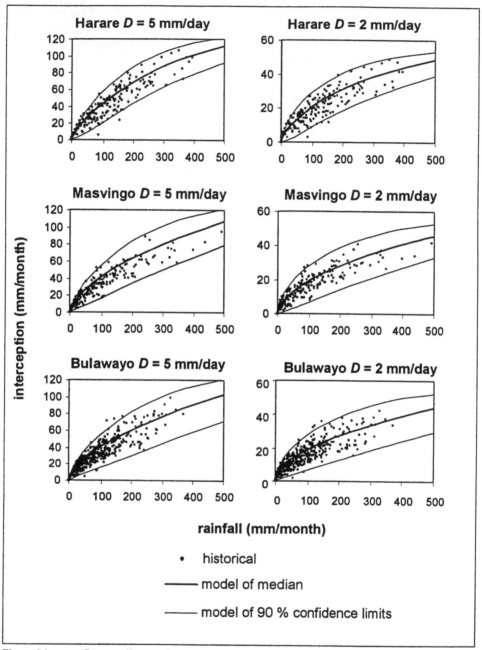

Figure 6.1   Scatter diagram between monthly total of rainfall and monthly total of daily interception. The lines give the monthly model for interception. *D* is the daily threshold. What is denoted as '90% confidence limits' are the equations that determine 95% and 5% of exceedance for the number of rain-days.

Because of Eq. 5.10:

$$E(n_r|n_m) = \frac{P_m}{E(P_r)} = \frac{P_m}{\beta} \qquad \text{for } P_r > 0 \text{ (mm/day)} \qquad \text{Eq. 6.8}$$

Combining Eqs. 6.3 - 6.8 results in an expression for the monthly interception

$$I_m = \frac{P_m}{\beta}\left\{ (-\beta - D)\exp\left(\frac{-D}{\beta}\right) + \beta + D\exp\left(\frac{-D}{\beta}\right) \right\}$$

$$= \frac{P_m}{\beta}\left\{ \beta\left(1 - \exp\left(\frac{-D}{\beta}\right)\right) \right\}$$

$$\text{Eq. 6.9}$$

which can be further simplified to

$$\boxed{I_m = P_m\left(1 - \exp\left(\frac{-D}{\beta}\right)\right)} \qquad \text{Eq. 6.10}$$

Substitution of $\beta = P_m(1-C)/(n_m p_{01})$ (Eq. 5.14) with $p_{01} = qP_m^r$ (Eq. 3.7) yields:

$$I_m = P_m\left(1 - \exp\left(\frac{-D * n_m q}{P_m^{1-r}(1-C)}\right)\right) \qquad \text{Eq. 6.11}$$

For Harare, C can be regarded constant with varying P_m. In cases where C may not be considered a constant, C in Eq. 6.11 should be substituted by Eq. 3.5, which yields:

$$I_m = P_m\left(1 - \exp\left(\frac{-D * n_m q}{P_m^{1-r}(1 - P_m^{v}(u - qP_m^{r-v}))}\right)\right)$$

$$= P_m\left(1 - \exp\left(\frac{-D * n_m q}{P_m^{1-r} - uP_m^{1-r+v} + qP_m}\right)\right)$$

$$\text{Eq. 6.12}$$

As mentioned in the previous chapter, β is a scale factor for the slope of the probability of exceedance of rainfall on a rain-day and is a function of the monthly rainfall P_m (see Eq. 5.14). The calibration coefficients calibrated are q, r and u, v or C.

Figure 6.1 shows that this equation is a good approximation of the relationship between monthly rainfall and the median interception derived from the daily rainfall values. The equations perform equally well for different threshold levels D.

Figure 6.2a shows that the equation for monthly median interception is not very sensitive to spatial differences in transition probability p_{01}. This means that the statistical characteristics of rainfall occurrence at a few locations in the region ($\sim$ 300 km) are sufficient to derive median interception.

Figure 6.2b shows that the mean interception per rain-day is very insensitive to spatial differences in the relationship between monthly rainfall and β. This implies that differences between locations in Eq. 6.10 mainly relate to differences in the relationship between the monthly rainfall and the number of rain-days.

Because β is the ratio of the monthly rainfall P_m to the number of rain-days n_r, Eq. 6.10 can be transformed to:

$$I_m = P_m \left(1 - \exp\left(\frac{-D * n_r}{P_m} \right) \right)$$

Eq. 6.13

As a consequence, for climates where the Markov process does not apply to the occurrence of rain-days, the interception equation can also be used if rainfall per rain-day agrees to an exponential distribution and if the relationship between the monthly rainfall and the mean number of rain-days is known.

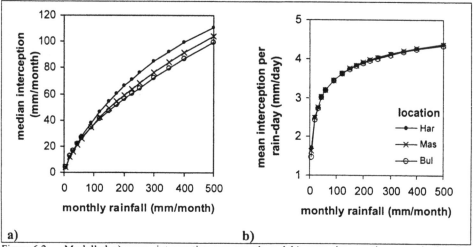

a)   b)

Figure 6.2   Modelled a) mean interception per month and b) mean interception per rain-day for Harare, Masvingo and Bulawayo for $D = 5$ mm/day.

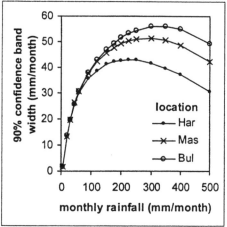

Figure 6.3   Band width of 90% confidence for modelled   monthly   interception   for   Harare, Masvingo and Bulawayo for $D = 5$ mm/day.

6.4   Confidence limits

For a given amount of monthly rainfall, the number of rain-days in a month is approximated by a normal probability density function of which mean and variance are known (Eqs. 4.2 – 4.3). Therefore, for any probability of exceedance the corresponding number of rain-days in the month can be derived. Consequently, instead of using the expected mean rainfall on rain-days, a β_x can be found that is the mean of the rainfall on rain-days with a certain probability of exceedance x.

$$\beta_x = \frac{n_m}{\text{invF}\left(x \middle| \text{N}\left(\dfrac{P_m(1-C)}{n_m p_{01}}, n_m \dfrac{p_{01}}{1-C}\left(1-\left(\dfrac{p_{01}}{1-C}\right)\dfrac{1+C}{1-C}\right)\right)\right)} \qquad \text{Eq. 6.14}$$

Substituting this β_x in Eq. 6.10 yields a confidence band around the monthly median interception. Figure 6.1 shows that almost all points of monthly rainfall and monthly interception derived from the daily model fall between the theoretically-derived 90% confidence limits.

Figure 6.3 illustrates that the modelled confidence band has a maximum width of between 200 and 350 mm/month rainfall, depending on the location. However, this this is not a very firm conclusion, because there are few data points for these extremely high rainfall amounts and the fit of the power function between monthly rainfall and transition probabilities is weak at such amounts of rainfall. The confidence band is widest for Bulawayo, which has the lowest values of p_{01}. The relative differences in the band width of interception are greater than in median interception.

6.5   Interception when monthly rainfall amount is very small

If the interception equation is suitable, it should also perform well at the extreme ends. It is clear that with very little monthly rainfall, all rainfall is intercepted. From Figure 6.1 it can be seen that at very little monthly rainfall the model considers almost all rainfall intercepted and that the median monthly interception tends towards the upper 90% confidence limit. In this subsection the analytical proof is given that at monthly rainfall of close to 0 mm/month all rainfall is indeed intercepted.

In equations this means that the following two equations should be proven for Eq. 6.12:

$$\lim_{P_m \downarrow 0} I_m = 0 \qquad \text{Eq. 6.15}$$

and

$$\lim_{P_m \downarrow 0} \frac{dI_m}{dP_m} = 1 \qquad \text{Eq. 6.16}$$

It is evident that Eq. 6.15 applies. To show that Eq. 6.16 also applies, Eq. 6.12 is differentiated.

$$\lim_{P_m \downarrow 0} \frac{\mathrm{d}I_m}{\mathrm{d}P_m}$$

$$= \lim_{P_m \downarrow 0} \left(\begin{array}{l} 1 - \exp\left(\dfrac{-Dn_m q}{P_m^{1-r} - uP_m^{1-r} + qP_m} \right) \\[3mm] -P_m * Dn_m q \left(P_m^{1-r} - uP_m^{1-r+v} + qP_m \right)^{-2} \\[3mm] * \left((1-r)P_m^{-r} - u(1-r+v)P_m^{-r+v} + q \right) \exp\left(\dfrac{-Dn_m q}{P_m^{1-r} - uP_m^{1-r} + qP_m} \right) \end{array} \right) \qquad \text{Eq. 6.17}$$

Because the first term is 1, to prove Eq. 6.16 the sum of the second and the third term should be 0. For notational convenience assume $x = (Dn_m q)/(P_m^{1-r} - uP_m^{1-r} + qP_m)$. For $r < 1$, which is always true, $x \rightarrow \infty$ when $P_m \rightarrow 0$. This means that the limit of the second term can be written as $\lim_{x \rightarrow \infty} \exp(-x)$, which is obviously 0.

It remains to be proven that the third term also approaches 0 when $P_m \rightarrow 0$. To determine the limit of the third term we use the same x for substitution and additionally assume $z = (P_m^{1-r} - uP_m^{1-r} + qP_m)$. For $r < 1$, $z \rightarrow 0$ when $P_m \rightarrow 0$. Substitution in Eq. 6.17 yields:

$$\lim_{P_m \downarrow 0,} \left(\begin{array}{l} -P_m * Dn_m q \left(P_m^{1-r} - uP_m^{1-r+v} + qP_m \right)^{-2} \\[3mm] * \left((1-r)P_m^{-r} - u(1-r+v)P_m^{-r+v} + q \right) \exp\left(\dfrac{-Dn_m q}{P_m^{1-r} - uP_m^{1-r} + qP_m} \right) \end{array} \right)$$

$$= \lim_{P_m \downarrow 0, x \rightarrow \infty, z \rightarrow 0} \left(\begin{array}{l} -Dn_m q \\[3mm] * \left(-rP_m^{1-r} - u(-r+v)P_m^{1-r+v} \right) \dfrac{\exp\left(\dfrac{-Dn_m q}{z} \right)}{z^2} - x \exp(-x) \end{array} \right) \qquad \text{Eq. 6.18}$$

It can be readily shown that $\lim_{y \rightarrow \infty} x \exp(-x) = 0$.

Let $y = \exp(x)$ then $x \exp(-x) = 1/y \ln(y)$. The limit for $\lim_{y \rightarrow \infty} 1/y * \ln(y)$ can be enclosed by two functions that become 0 (see mathematical handbooks, e.g. Almering, 1988):

$$0 < \frac{\ln(y)}{y} = \frac{\int_1^x \dfrac{\mathrm{d}t}{t}}{y} \leq \frac{\int_1^x \dfrac{\mathrm{d}t}{\sqrt{t}}}{y} = \frac{2}{\sqrt{y}} - \frac{2}{y} \qquad \text{Eq. 6.19}$$

Hence it remains to be shown that

$$\lim_{P_m \downarrow 0, z \rightarrow 0} \left(-Dn_m q * \left(-rP_m^{1-r} - u(-r+v)P_m^{1-r+v} \right) \dfrac{\exp\left(\dfrac{-Dn_m q}{z} \right)}{z^2} \right) = 0 \qquad \text{Eq. 6.20}$$

It is clear that $\lim_{P_m \downarrow 0} \left(-D n_m q * \left(-r P_m^{1-r} - u(-r+v) P_m^{1-r+v} \right) \right) = 0$ for $1-r > 0$ and for $1-r+v$

> 0, which are both always true. Subsequently it needs to be shown that the remaining factor is finite:

$$\lim_{z \to 0} \frac{\exp\left(\dfrac{-D n_m q}{z} \right)}{z^2} = \lim_{w \to \infty} \frac{w^b}{a^w} \qquad \text{Eq. 6.21}$$

for $w = 1/z$, $a = \exp(D n_m q)$, $b = 2$.

Let $x = 1/b * z * \ln(a)$. For any $a > 1$ and $b > 0$ and under the conditions that $1-r > 0$ and $1-r-v > 0$, it follows that (Almering, 1988):

$$\lim_{x \to \infty} x \exp(-x) = \lim_{z \to \infty} \frac{z \ln(a)}{b} \exp\left(-\frac{z \ln(a)}{b} \right) = 0 \qquad \text{Eq. 6.22}$$

The condition $1-r-v > 0$ is the most restrictive condition set to r and v. For the daily rainfall series tested in Zimbabwe, $1-r-v > 0.15$. The station of Chinhoyi has the smallest value of $1-r-v$ (0.16), followed by Harare (0.21).

Eq. 6.22 is used for the following proof:

$$\lim_{w \to \infty} \frac{w^b}{a^w} = \lim_{w \to \infty} \left(\frac{\dfrac{w \ln(a)}{b}}{a^{\frac{w}{b}}} \right)^b \left(\frac{b}{\ln(a)} \right)^b = 0 * \left(\frac{b}{\ln(a)} \right)^b = 0 \qquad \text{Eq. 6.23}$$

Herewith is proven that $\lim_{P_m \downarrow 0} \dfrac{dI_m}{dP_m} = 1$.

6.6 Expressions for effective rainfall at daily and monthly time steps

Effective rainfall is the part of rainfall that is not intercepted and subsequently feeds transpiration and runoff processes. In the previous paragraph an expression was found for the monthly interception. The median of rainfall on rain-days $M(P_{eff,r})$ is expressed as β. Hence the following relation can be found for the median of effective rainfall on rain-days:

$$M(P_{eff,r}) = M(P_r) - \frac{I_m}{E(n_r|n_m)}$$

$$= \beta - \frac{P_m \left(1 - \exp\left(\dfrac{-D}{\beta} \right) \right)}{P_m / \beta} \qquad \text{Eq. 6.24}$$

which can be further simplified to:

$$\boxed{ M(P_{eff,r}) = \beta \exp\left(\frac{-D}{\beta} \right) } \qquad \text{Eq. 6.25}$$

Thus, simply knowing the mean rainfall on rain-days and the interception threshold is sufficient to estimate the median of effective rainfall on rain-days.

At monthly time steps the expression for effective rainfall ($P_{eff,m} = P_m - I_m$) immediately follows from the equation for monthly interception Eq. 6.10:

$$P_{eff,m} = P_m \exp\left(\frac{-D}{\beta}\right)$$

Eq. 6.26

6.7   Other monthly interception models

Zeng

After the analytical equations for monthly interception had been developed (De Groen, 1999), Zeng, Shuttleworth & Gash (2000) published an alternative equation for monthly interception, also taking account of variability of rainfall within the month. Instead of using a daily threshold, Zeng et al. (2000) used the Rutter interception model as the basis (original in Rutter et al., 1971). Zeng used hourly data to fit the model, and therefore the method is not suitable for conditions in which only daily data are available. However, it is worth comparing this technique with the one used in this dissertation.

The Rutter model is used in many GCM's for land surface parameterisation, some of which have modifications to take account of spatial variability of rainfall, when applied to model grid boxes (Zeng et al., 2000). However, these parameterisations do not properly account for the temporal variability of rainfall. Therefore Zeng et al. developed an alternative parameterisation.

The Rutter model is a physically-based numerical point interception model. This model has been the basis of numerous subsequent analytical or semi-analytical models. A running water balance is maintained for the canopy storage. As long as the canopy storage is at less than maximum, the evaporation is assumed to be proportional to the storage. When the canopy storage is full, the evaporation is at the potential rate, as can be computed with the Penman-Monteith equation (see Chapter 10). During a storm the canopy may not reach saturation, in which case the potential evaporation is not reached. As soon as the storm has stopped evaporation is in any case less than the potential. Thus:

$$I = \frac{D_{act}}{D_v} I_{pot} \qquad\qquad D_{act} \le D_v \qquad \text{(mm/day)} \qquad \text{Eq. 6.27}$$

where

D_{act}	is actual water storage in canopy	(mm)
D_v	maximum canopy storage	(mm)
I_{pot}	potential evaporation from a wet canopy, following the Penman-Monteith approach (Appendix C)	(mm/day)

The approach of the Rutter model to interception shows great similarity with the spells approach to transpiration in this dissertation. In the Rutter model, potential evaporation is assumed to be the same during storms as it is in between storm periods, while in reality during periods between storms potential evaporation is higher than during the storms. (Discussion follows in Chapter 10.) Moreover, Zeng et al. do not implement a diurnal variation in potential evaporation.

As is done in this dissertation, Zeng et al. analytically derive monthly interception from the functions that are the basis of a stochastic rainfall model. The model is continuous. To describe temporal variability of rainfall Zeng et al. assume exponential probability density functions for storm intensity i, storm duration t_v and inter-storm break time t_b, respectively as:

$$f_i(i) = \frac{1}{i_m}\exp\left(\frac{-i}{i_m}\right) \qquad \text{Eq. 6.28}$$

$$f_r(t_r) = \frac{1}{\tau_r}\exp\left(\frac{-t}{\tau_r}\right) \qquad \text{Eq. 6.29}$$

$$f_b(t_b) = \frac{1}{\tau_b}\exp\left(\frac{-t}{\tau_b}\right) \qquad \text{Eq. 6.30}$$

where

i_m	is mean storm intensity	(mm/h)
τ_r	mean storm duration	(h)
τ_b	mean inter-storm break time	(h)

To maintain analytical tractability, the distributions are assumed independent of each other. Additionally, a very useful time scale in the derivation is the time it takes to evaporate a saturated canopy at its potential rate:

$$\tau_0 = \frac{D_v}{I_{pot,h}} \qquad \text{(mm/h)} \qquad \text{Eq. 6.31}$$

The mean storm intensity i_m and the mean storm duration τ_r are taken as mean values: the results are found to be insensitive to the making of a distinction in these values between months. Thus the inter-storm break time τ_b is the only parameter differing per month. This parameter varies linearly with the reciprocal of monthly rainfall:

$$\tau_b = \frac{n_m \Delta t_d}{P_m i_m \tau_r} - \tau_r \qquad \text{(mm/h)} \qquad \text{Eq. 6.32}$$

Using these assumptions, Zeng et al. (2000) arrive at four versions of monthly interception equations. The exact solution reads:

$$I_{mz} = \frac{P_m}{i_m * \Delta t_d}\left\{\alpha_1 + \frac{\tau_0}{\tau_r}\left[\alpha_2\left(1+\frac{\tau_0}{\tau_b}\right)^{-1} - \alpha_3\right]\right\} \qquad \text{(mm/month)} \qquad \text{Eq. 6.33}$$

where

$$\alpha_1 = 1 - \frac{I_{pot}\tau_r}{i_m \tau_0} + \frac{\alpha_3 \tau_r^2}{\tau_0^2} \qquad \text{(-)} \qquad \text{Eq. 6.34}$$

$$\alpha_2 = 1 - \frac{2\alpha_3 \tau_r}{\tau_0} \qquad \text{(-)} \qquad \text{Eq. 6.35}$$

$$\alpha_3 = \frac{I_{pot}}{2i_m}\ln\left(\frac{\tau_r i_m}{\tau_0 I_{pot}}\right) \qquad \text{(-)} \qquad \text{Eq. 6.36}$$

and $\Delta t_d = 24$ h/day.

As a first approximation α_1 is assumed a constant, together with

$$\alpha_4 = \alpha_2 \left(1 + \frac{\tau_0}{\mu_{\tau_b}} \right)^{-1} - \alpha_3 \qquad\qquad (\text{-}) \qquad\qquad \text{Eq. 6.37}$$

Thereby the first approximation becomes

$$I_{mz1} = \frac{P_m}{i_m \Delta t_d} \left\{ \alpha_1 + \alpha_4 \frac{\tau_0}{\tau_r} \right\} \qquad\qquad (\text{mm/month}) \qquad\qquad \text{Eq. 6.38}$$

In a second approximation, storms are assumed sufficiently strong to saturate the canopy instantly:

$$I_{mz2} = \frac{P_m}{i_m \Delta t_d} \left\{ 1 + \frac{\tau_0}{\tau_r} \left(1 + \frac{\tau_0}{\tau_b} \right)^{-1} \right\} \qquad\qquad (\text{mm/month}) \qquad\qquad \text{Eq. 6.39}$$

In the third approximation the inter-storm break times are assumed long enough for all the water to be evaporated before the next storm arrives:

$$I_{mz3} = \frac{P_m}{i_m \Delta t_d} \left\{ 1 + \frac{\tau_0}{\tau_r} \right\} \qquad\qquad (\text{mm/month}) \qquad\qquad \text{Eq. 6.40}$$

How does the method used by Zeng et al. compare to modelling with the Markov chain, as is done in this dissertation? In the Markov chain approach storms occurring on the same day are aggregated. The daily time step has a physical meaning:
•   In a climate dominated by convective storms normally only one storm per day occurs, usually at the end of the day when Convectional Available Potential Energy is sufficient. As a consequence, the inter-storm break time τ_b has a probability density function that shows local maxima at multiples of 24- τ_r hours.
•   Evaporation of interception has a diurnal variation. Zeng et al. ignore this aspect of variability within a day.

For verification, Zeng et al. use data from the ARME experimental site in the Amazon and from the station Les Landes in France. In both cases the mean inter-storm break time τ_b is in the order of 30 hours, thus significantly longer than 24 hours, which means that most often only one storm occurs in a 24-hour period. If only one storm a day occurs, daily rainfall is the product of storm intensity and storm duration, which both agree to an exponential probability density function. The product of two parameters which both agree to an independent exponential probability distribution function has a probability density function that is similar to an exponential distribution. Therefore, the probability density function of rain on rain-days in the model by Zeng et al. approaches an exponential distribution, as with the model in this dissertation. However, in the model by Zeng et al. the scale factor cannot be derived in a straightforward way.

The Markov chain gives an exponential distribution for lengths of dry spells. This is similar to the exponential distribution of storm break times in Zengs model. The mean storm duration is in the order of 2 hours, after which an inter-storm break time again follows, drawn from an exponential probability density function.

Zeng et al. derive an equation for monthly interception that is proportional to monthly rainfall (I_{mz4}), or almost proportional (I_{mz}, I_{mz1}, I_{mz2}, I_{mz3}). Figure 6.4 shows the different models by Zeng et al., applied on parameters from Les Landes, France, as no hourly data from Zimbabwe were available. That the models are less concave than the model proposed in this dissertation is not the result of the analytical derivation as such, but rather of the assumption that the storm inter-arrival time ($\tau_b + \tau_r$) is proportional to the reciprocal value of the monthly rainfall. In Chapter 2 it is shown that this is not necessarily valid for all countries where rainfall occurrence follows a Markov process. Logically, if the monthly rainfall amount is large, then the probability of a big storm in this month is large as well. Thus the number of storms in a month is a concave function of monthly rainfall. Consequently the storm inter-arrival time has a slope that is less than the reciprocal value of monthly rainfall. Zeng et al.'s method agrees very well with monthly totals of the Rutter model output, using hourly data from two experimental sites. However, for cases where rainfall occurrence follows the Markov process the method is not suitable even if hourly rainfall data are available. Zeng et al. do not discuss how spatially homogeneous the parameters i_m, τ_r, τ_b are, and thus what density of stations with records of hourly rainfall is necessary.

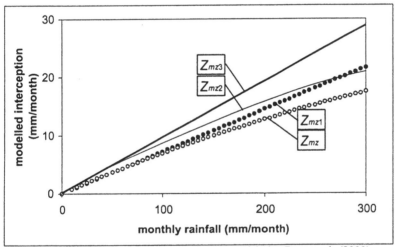

Figure 6.4   Monthly interception according to models by Zeng et al. (2000) as a function of monthly rainfall, on the basis of parameters of Les Landes, France ($\tau_r = 2.5$ h, $\mu(\tau_b) = 30.7$ h, $i_m = 1.0$ mm/h, $D_v = 0.56$ mm, $I_{pot.h} = 0.17$ mm/h, thus $\tau_0 = 3.3$ h, $\alpha_1 = 1.10$, $\alpha_2 = 0.67$, $\alpha_3 = 0.12$, $\alpha_4 = 0.48$).

Pitman

Pitman (1973) used one daily interception model for many stations in South Africa. This daily model uses the same assumptions as are used in this dissertation. From the monthly sums of the output, he empirically derived a monthly model for interception, which reads:

$$I_m = 13.08 * D^{1.14} * \left(1 - \exp\left(P_m \left(0.00099 * D^{0.75} - 0.011\right)\right)\right) \qquad \text{Eq. 6.41}$$

The values of the model parameters have no physical meaning and the units, e.g. $(mm/day)^{1.14}$, are physically meaningless. The equation suggests that the relationship between the monthly rainfall and the interception threshold is homogeneous for the whole of South Africa, which is extremely unrealistic, because the daily variability of rainfall is very different in the various regions (Zucchini et al., 1992). However, Figure 6.5 shows that Pitman's empirical equation shows a similar pattern to that of the model in this dissertation.

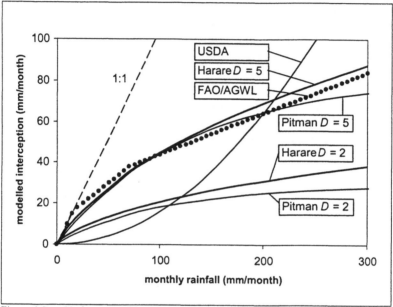

Figure 6.5   Various monthly interception models: Pitman, USDA, FAO/AGWL, and interception in Harare according to this dissertation.

USDA

The FAO model CROPWAT (Clarke et al., 1998) is generally applied to estimate the performance of different cropping patterns and water application scheduling and to evaluate rainfed agricultural production. It also uses monthly rainfall input data. Effective rainfall is default computed using the method of the USDA Soil Conservation Service. Assuming that interception is the difference between gross rainfall and effective rainfall, the USDA approach reads:

$$I_m = P_m - P_{eff,m} = \frac{0.2}{125} * P_m^{\ 2} \qquad \text{for } P_m < 250 \text{ mm/month} \qquad \text{Eq. 6.42}$$

$$I_m = P_m - P_{eff,m} = 0.9 * P_m - 125 \qquad \text{for } P_m > 250 \text{ mm/month} \qquad \text{Eq. 6.43}$$

which does not reflect the daily model in this dissertation at all, and thus is probably based on different assumptions about daily interception.

FAO/AGWL

CROPWAT also offers the possibility of using the linear FAO/AGWL equation:

$$I_m = \text{Min}(0.4 * P_m + 10, P_m) \qquad\qquad \text{for } P_m < 70 \text{ mm/month} \quad \text{Eq. 6.44}$$

$$I_m = 0.2 * P_m + 24 \qquad\qquad \text{for } P_m > 70 \text{ mm/month} \quad \text{Eq. 6.45}$$

Figure 6.5 shows that this empirical model fits reasonably well with the model of Harare. Yet the FAO/AGWL equation does not have a variable for the interception threshold D, which is dependent on land cover. The CROPWAT developers seem to realise this deficiency, because the possibility of adjusting all constants to local conditions is offered.

Schreiber

Schreiber (1904; see also Dooge, 1997) did not derive a monthly interception equation, but empirically derived a relationship between long-term rain totals for different locations in Central Europe to the long-term runoff as derived from water balances. The similarity of Schreiber's model of evaporation with the interception equation of this dissertation is remarkable. Schreiber found that

$$E = P\left(1 - \exp\left(-\frac{E_{pot}}{P}\right)\right) \qquad\qquad \text{for } P_m < 70 \text{ mm/month} \quad \text{Eq. 6.46}$$

where each of the parameters is a long term average of

E	actual evaporation	(mm/day)
P	rainfall	(mm/day)
E_{pot}	potential evaporation	(mm/day)

6.8   Conclusions

It has been shown in this chapter that an analytical relationship between monthly rainfall and monthly interception can be derived from the Markov process for rainfall occurrence and the exponential probability density function for rainfall amounts on rain-days. The model compares favourably with existing monthly interception models, because of the transparent link with the daily interception threshold.

For climates where the Markov process does not apply to the occurrence of rain-days, the interception equation can also be used if rainfall per rain-day agrees to an exponential distribution and if the relationship between the monthly rainfall and the mean number of rain-days is known.

The monthly interception is not very sensitive to the spatial differences in transition probability p_{01}. Therefore statistical characteristics of rainfall occurrence at a few locations in the region (~ 300 km) are sufficient to derive the necessary parameters.

7 Transpiration - the Spells Approach

7.1 Introduction

Modelling transpiration at monthly time steps is more complicated than modelling interception, because transpiration depends on the soil moisture storage. On the other hand, as a result of the storage component, transpiration has a longer process time scale than interception. Depending on the ratio of storage to potential transpiration, the time scale is in the order of 10 days to a month. Incorporating statistics of daily data in analytical equations of monthly models requires incorporating the fluctuation of soil moisture storage within a month. This is discussed in this Chapter.

Readers who are interested in an easily applicable monthly model for transpiration and not so much in the derivation are being recommended to move directly to Chapter 8.

7.2 Daily model

The model that is presented in this Chapter is based on the principles of a daily model, with which it will be compared later. For this daily model some assumptions on the functioning of the soil water reservoir are used. These assumptions will first be mentioned and then discussed.

- Effective daily rainfall is the gross rainfall minus the daily interception as defined in Chapter 6.
- Transpiration is equal to potential transpiration, unless the available soil water content is below a certain limit. This limit S_b is usually 0.5-0.8 of the maximum available soil moisture content S_{max}, depending on the soil and the vegetation. Below that limit the transpiration decreases proportionally to the available soil water content.
- Hortonian overland flow is not a limitation to soil moisture replenishment.
- Infiltration is assumed to benefit the soil water content homogeneously.
- Only when the soil water storage is saturated does rainfall recharge groundwater or contribute to overland flow.
- The soil moisture is not fed by groundwater in the rainy season.

In this Section the different assumptions are elaborated and argued, in particular for Zimbabwe.

It was mentioned in the previous Chapter that surface water retention can be greater than what is defined as interception. However, here infiltration to the soil moisture on a certain day is assumed to be the difference between the amount of rain and the interception.

Box 7-A   Taking account of Hortonian overland flow

In locations where the infiltration capacity can be said to be a limiting factor on a daily time scale, overland flow can be determined on a monthly time scale by a similar approach to the one applied in the previous Chapter for interception. Where interception is all rainfall that is less than a certain daily threshold D, overland flow is all rainfall that exceeds a certain daily threshold D_{hort} (mm/day). In analogy with Eq. 6.10, the equation for monthly Hortonian overland flow $R_{hort,m}$ yields:

$$R_{hort,m} = P_m \exp\left(\frac{-D_{hort}}{\beta}\right)$$ (mm/month)   Eq. 7a.1

However, because the probabilities of daily rainfall exceeding D_{hort} are far smaller than those of exceeding D, the confidence band is much wider than in the case of interception (see Eq. 6.13). If it is chosen not to neglect Hortonian overland flow, all equations in this Chapter and the following Chapter remain valid. It is only that the meaning of the parameter of effective monthly rainfall $P_{eff,m}$ changes from $P_{eff,m} = P_m - I_m$ to

$$P_{eff,m} = P_m - I_m - R_{hort,m}$$

$$= P_m\left(\exp\left(\frac{-D}{\beta}\right) - \exp\left(\frac{-D_{hort}}{\beta}\right)\right)$$ (mm/month)   Eq. 7a.2

and the meaning of the expected effective daily rainfall changes from $M(P_{eff,r}) = \beta \exp(-D/\beta)$ (Eq. 6.24) to:

$$M(P_{eff,r}) = \beta\left(\exp(-D/\beta) - \exp(-D_{hort}/\beta)\right)$$ (mm/day)   Eq. 7a.3

Researchers are in broad agreement that soil moisture constrained transpiration is directly proportional to the available soil moisture (see textbooks, e.g. Shuttleworth, 1997; Dingman, 1998).[27] Equations relating soil moisture and transpiration have the form (see Figure 7.1):

$$T_{act} = T_{pot} * \min\left(\frac{S}{S_b}, 1\right)$$ Eq. 7.1

where

S is the available soil moisture content, which is the actual soil moisture content minus the wilting point   (mm)

S_b the available soil moisture content below which transpiration is soil moisture constrained   (mm)

S_b is typically 50 to 80% of the maximum available moisture content in the root zone S_{max} (Shuttleworth, 1997).[28] The ratio depends on atmospheric demand and on vegetation. With any soil moisture content between S_b and S_{max}, the soil will evaporate at a rate T_{pot}.

---

[27] When modelling the change in transpiration in response to soil water status it is convenient to work in terms of the soil water content, but plants are sensitive to the soil water potential ψ_s, the energy necessary to bring the water up for transpiration. Some of the variability between S and T is caused by the variability between ψ_s and S for different soils (Shuttleworth, 1997).

[28] In particular for annual crops the rooting depth changes as the vegetation grows. Also, in a mixture of crops each can have a different rooting depth. The most realistic of soil moisture restriction models simulate plant extraction from a series of moisture stores vertically above each other (Shuttleworth, 1997).

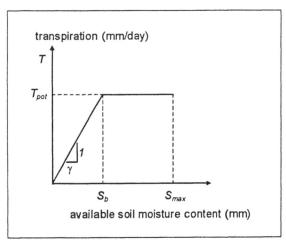

Figure 7.1   Typical form of relationship between soil
moisture content and actual transpiration.

Penman has been the first to describe a linear relationship between soil moisture content and transpiration, where evaporation was assumed potential if the soil moisture deficit was not less than a so-called 'root constant' (1949, see Alley, 1984). Thornthwaite & Mather (1955 & 1957) simplified Penman's findings and described a proportional relationship between S and E in order to derive monthly models for evaporation. Thus, Thornthwaite & Mather assumed that $S_b = S_{max}$. Up to the present date, most monthly models are based on this same assumption of proportionality. Therefore, in Chapter 8 the Thornthwaite-Mather model is compared with the model in this dissertation.

On a daily basis it is assumed that the effective rainfall infiltrates, except when the soil is fully saturated. This implies that the infiltration capacity of the soil is never a limiting factor. In other words, Hortonian overland flow is neglected. For the purpose of deriving monthly equations for transpiration, Hortonian overland flow is not really relevant, for the following reasons:

• The application of the monthly transpiration model that has the smallest spatial scale is the application for crop yield assessment. In fields with crops infiltration capacities are not very low and therefore Hortonian overland flow is not really important.

• The other model applications are for water resources assessment on a catchment scale > 100 (km)². In this case the overland flow is of little relevance because most Hortonian overland flow infiltrates in close vicinity of the place where the rain fell.

As regards Zimbabwe, Wolski (1999) quotes Bell et al. (1987) that for Chizengeni catchment interfluves infiltration capacity is 230 mm/h. McCartney (1998) found values for the Grasslands catchment of 585 mm/h. In valley bottoms in Zimbabwe, Hortonian overland flow may be locally important. For central Ivory Coast, Van de Giesen et al. (2000) found that the Hortonian overland flow reduced by some 50% comparing slopes with a length of 12 m to those of 1.25 m. For locations where Hortonian overland flow cannot be neglected, the equations in this and the following Chapter can easily be adopted to take account of it (see Box 7-A).

Preferential flow paths mainly occur in soils that are dry. In general, with regard to percolation, daily soil water balance models do not make a distinction between different initial states of the soil moisture. This implies that if preferential flow occurs, it is assumed not to affect the linear relationship between average soil moisture storage and transpiration.

It is assumed that no percolation to the groundwater takes place as long as the soil is not saturated. For water balance models at monthly time steps this is an established approach. Both the Thornthwaite-Mather and the Palmer models use this assumption (Thornthwaite & Mather, 1957; Palmer, 1965, see Alley, 1984) and they are used up to the present time. Recharge rates in monthly models will be underestimated if recharge takes place only on saturation of the soil (Morton, 1983; Rushton & Ward, 1979). However, Steenhuis & Van der Molen (1986) determined the recharge at Long Island (New York State, USA) accurately with the use of the Thornthwaite-Mather approach at daily time steps. Good results have also been reported for the Ivory Coast, Rwanda and Zimbabwe (personal communication, 2000, M. Masiyandima, IWMI). Wolski (1999) used the Thornthwaite-Mather method to estimate recharge in the Lui Luena triangle in Zambia. The values for recharge, for a realistic range of root zone depth, were lower than values determined with the chloride method and higher than minimum values determined with water balance methods.

Other models regard recharge differently. The daily HBV model (Lindström et al., 1997) and the model by Wilby et al. (1994) allow a fixed proportion of the precipitation to bypass the soil horizon during periods of soil moisture deficit. Wilby et al. claim that such an approach better recognises groundwater and river flow fluctuation during summer in Great Britain. In semi-arid areas, recharge in the dry season (winter) is negligible, therefore Wilby et al.'s argument to use recharge proportional to effective rainfall is not applicable.

Considering the lack of detailed soil water storage parameters and scarce direct measurements of recharge to support water resources assessment models, it is justifiable to accept the threshold concept that recharge, like runoff, does not occur until the soil moisture content is at maximum capacity.

The remaining assumption to be argued is that the soil moisture is not fed by groundwater in the rainy season. McCartney (1998) made detailed measurements of soil moisture contents and groundwater levels at Grasslands headwater catchment with a dambo, at the end of the dry season after a relatively wet year (08/09/1996 to 25/10/1996). Earlier research by Bullock (1992a,b) on dambos in Zimbabwe showed that dry season transpiration is relatively high there. In Grasslands the groundwater levels ranged from 1 to 4 meters below the surface, fairly high. The total change in soil water during the dry season amounted to 1.5 mm/day. Comparison of the changes in groundwater storage and soil moisture depletion indicated that 85% of the change in subsurface water was due to a change in the unsaturated zone. Assuming the depletion of the groundwater can be fully attributed to transpiration gives a high estimate of capillary rise. Even in a case where all saturation depletion is attributed to evaporation, this is only 15% of the transpiration in an area with a relatively high water table at the end of the dry season. The contribution of groundwater to the soil moisture in the rainy season is therefore probably negligible in most locations.

The NAM model has a component for capillary flux (CAFLUX) which feeds water from the groundwater back to the soil moisture (Refsgaard & Knudsen, 1996). The HBV model has a component for capillary flux (CF) which feeds water back from the upper to the lower response box (Lindström et al., 1997). However, from the discussion on the observations by McCartney it is here concluded that feedback of water from the groundwater to the soil moisture is negligible in Zimbabwe, at least in the rainy season. This simplifies the monthly computation of transpiration considerably.

7.3 Disaggregation of monthly rainfall

Chapter 3 and 4 provided key parameters to describe the variability of rainfall within a month. In this Section the disaggregation of monthly rainfall is presented, which is used to model the flux of water to the soil moisture.

Within the month the following parameters are constant and are used as input parameters:
- the conditional probabilities p_{01} and p_{11} as a function of monthly rainfall (-),
- the daily interception threshold D (mm/day).

As discussed in Section 4.5, the theoretical 'average month' has λ pairs of a dry and a wet spell, which each have the expected length. The basic idea is that the monthly rainfall is disaggregated over the month to the wet spells in an 'average month' and that the rainfall is homogeneous within these wet spells.

The variance of the average daily rainfall in an average wet spell is considerable. The coefficient of variation ranges from 0.7 with little monthly rainfall to 0.2 with a large monthly rainfall amount (see Box 7-A). However, it is here assumed that rainfall is uniformly distributed over a wet spell. The intensity of rainfall is taken as constant over the wet spell and equal to the expected rainfall on rain-days β (mm/day). With this approach not only is the variability of rainfall on different rain-days in a wet spell neglected, but also the diurnal variability, with rain occurring most often at the end of the day. The soil moisture storage smoothes these variabilities out to some extent.

To recapitulate, in Chapters 4, 5 and 6 several parameters of interest have been estimated as a function of the monthly rainfall:
- expected number of pairs of a dry and wet spell in the month $E(N_{pairs}) = \lambda = p_{01}/(1 - p_{11} + p_{01})*(1-p_{11})*n_{m}$, see Eq. 4.29,
- expected length of a dry spell, $E(n_{dry}) = 1/p_{01}* \Delta t_d$, see Eq. 4.11,
- expected length of a wet spell, $E(n_{wet}) = 1/(1-p_{11}) * \Delta t_d$, see Eq. 4.16,
- expected effective rainfall $E(P_{r.eff}) = \beta \exp(-D/\beta)$ on rain-days, with $\beta = P_m*(1 - p_{11} + p_{01})/(n_m p_{01})$ as the expected gross rainfall on a rain-day (mm/day, see Eqs. 6.25 and 6.14).

The Markov property is based on discrete time steps of a day. However, expected lengths of wet and dry spells have been derived as the first moment of the discrete probability distribution function of spell lengths. As a consequence the expected lengths are not necessarily integers.

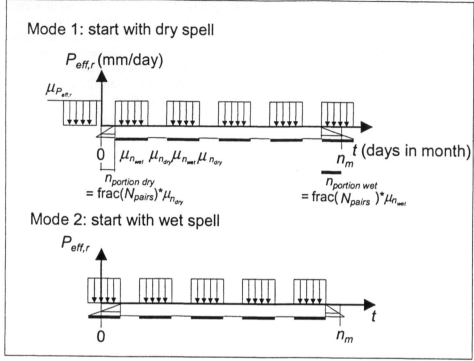

Figure 7.2   Schematisation of influx of rain into soil moisture during the month.

In Eq. 4.28 it was shown that the theoretical case of having only spells of expected length would yield a month which, if it starts with part of a dry or a wet spell, ends with part of the other type of spell. This results in two modes: one starting with a wet spell, the other starting with a dry spell. To estimate monthly transpiration, the average of the two modes is used. Figure 7.2 illustrates the two modes in a schematisation.

The dry spell that is not of the expected length relates to a dry spell of the expected length at the same ratio as the wet spell that is not of the expected length relates to the wet spell of expected length. See the triangles in Figure 7.2. The fraction relates to the number of pairs N_{pairs}, which is not necessarily an integer:

$$\text{frac}(N_{pairs}) = \frac{n_{portion\,dry}}{\mu_{n_{dry}}} = \frac{n_{portion\,wet}}{\mu_{n_{wet}}}$$ Eq. 7.2

where frac(N_{pairs}) is the fraction part of N_{pairs}. For example, if N_{pairs} = 4.3, frac(N_{pairs}) = 0.3.

Comparison with other models

This Section described a distribution of rainfall within the month that is the one most probable. As mentioned in Section 2.3, Mulligan & Reaney (1999) also used Markov processes to disaggregate monthly rainfall. However, they used a stochastically-generated disaggregation, which is non-unique. Pitman (1973, see also Hughes, 1997) developed a water resources model with monthly in- and output but with numerical time steps within the month, similar to the spells approach. In Pitman's model the cumulative rainfall distribution within the month is assumed to be S-shaped, which implies that the intensity of the rainfall is low at the end and start of the month and high in the middle. The S-shape has an equal maximum deviation below and above the cumulative distribution of a uniform rate, which is a straight line.

Pitman plotted the cumulative rainfall distribution within the month for many rainfall stations with daily data and determined the maximum deviation below and above the uniform rate line. In this way he derived an empirical relationship between the monthly precipitation and the sum of the two deviations for South Africa. The slope decreases with the monthly rainfall (as is the case with the interception equation in this dissertation). The S-shaped rainfall distribution is finally attributed to numerical time steps within the month. For South Africa four time steps appeared sufficient.

The disaggregation of monthly rainfall proposed in this dissertation is more representative of the within month rainfall distribution in terms of expected lengths of dry and wet spells than the methods of Mulligan & Reaney (1999) or Pitman (1973). However, this technique does not cater for the possibility of high rainfall intensities. Pitman considers this, although he takes an average maximum intensity and the scatter is quite large. Mulligan & Reaney occasionally use rainfall concentrated in a few days and occasionally a more uniform rate, depending on what their stochastic model generates.

7.4 Transpiration during dry and wet spells

Transpiration during a dry spell

If the soil moisture content is below S_b, thus $S<S_b$, transpiration is moisture constrained and proportional to the soil moisture content (Eq. 7.1). For given soil conditions and monthly rainfall:

$$T = -\frac{dS}{dt} = T_{pot}\frac{S}{S_b} \qquad \text{for } S < S_b \qquad \text{(mm/day)} \qquad \text{Eq. 7.3}$$

For notational convenience a new parameter γ is introduced, which is a time scale for the depletion of soil moisture below S_b. It is the time it would take for a soil moisture amount S_b to evaporate, if transpiration were to continue at a potential rate (see Figure 7.1).

$$\gamma = \frac{S_b}{T_{pot}} \qquad \text{(days)} \qquad \text{Eq. 7.4}$$

Consequently, Eq. 7.3 can be rewritten:

$$T = \frac{S}{\gamma} \qquad \text{for } S < S_b \qquad \text{(mm/day)} \qquad \text{Eq. 7.5}$$

The time is measured from the start of the month where
$t_{dry,i}$ is time at start of the *i*th dry spell in a month   (days)

Box 7-B   The variance of average daily rain over a mean wet spell

In the spells model for transpiration it is assumed that each wet spell is of the same length and has the same average daily rainfall. In this box it is explored what the variance is of the average daily rainfall, given that the wet spell is of the expected length.

The expected rainfall on a rain-day β is derived as if it were the median of the monthly averaged rainfall on rain-days. The variance of rainfall on rain-days $Var(P_r) = \beta^2$ (Eq. 6.6) applies to an infinite number of rain-days. However, for a certain wet spell with a length n_{wet}, the variance of the average rainfall on rain-days in the spell also depends on the number of rain-days in the spell, as is normally the case in statistics of samples. Hence for the variance of the average over n_{wet} rain-days:

$$\mathrm{Var}\left(\overline{P_r}\big|n_{wet}\right) = \frac{\mathrm{Var}(P_r)}{n_{wet}} \qquad\qquad \text{Eq. 7b.1}$$

Thus, for a mean spell length $\mu_{n_{wet}} = 1/(1 - p_{11})$ (Eq. 4.16) and $Var(P_r) = \beta^2$ yields:

$$\mathrm{Var}\left(\overline{P_r}\big|\mu_{n_{wet}}\right) = \beta^2(1 - p_{11}) \qquad\qquad \text{Eq. 7b.2}$$

The expectation of the average of daily rainfall in an average wet spell is the same as the mean of daily rainfall on rain-days $\mathrm{E}\left(\overline{P_r}\big|\mu_{n_{wet}}\right) = \mathrm{E}(P_r) = \beta$. Therefore the coefficient of variation of the mean daily rainfall in a wet spell simply yields:

$$\mathrm{C_v}\left(\overline{P_r}\big|\mu_{n_{wet}}\right) = \frac{\sigma\left(\overline{P_r}\big|\mu_{n_{wet}}\right)}{\mathrm{E}\left(\overline{P_r}\big|\mu_{n_{wet}}\right)} = \sqrt{1 - p_{11}} \qquad\qquad \text{Eq. 7b.3}$$

In Figure 7B.1 this ratio is shown as a function of monthly rainfall. In doing so it is disregarded that monthly rainfall is the starting point and thus $\mu_{n_w} * \sum_0^w \left(\overline{P_r}\big|\mu_{n_{wet}}\right) = P_m$.

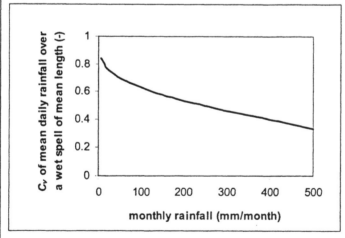

Figure 7B.1   Coefficient of variability C_v of average daily rainfall during wet spell of expected length.

For a particular soil moisture content at the start of the dry spell $S_{start,dry,i} < S_b$, interpolation of Eq. 7.3 yields:

$$S(t) = S_{start,dry,i} * \exp\left(-\frac{t - t_{dry,i}}{\gamma}\right)$$
Eq. 7.6

Consequently:

$$T(t) = \frac{S_{start,dry,i}}{\gamma} * \exp\left(-\frac{t - t_{dry,i}}{\gamma}\right)$$
Eq. 7.8

Therefore, the total transpiration during the dry spell reads:

$$\int_{t_{dry,i}}^{t_{dry,i}+\mu_{n_{dry}}} T(t)\, dt = S_{start,dry,i}\left(1 - \exp\left(-\frac{\mu_{n_{dry}}}{\gamma}\right)\right) \qquad \text{for } S_{start,dry,i} < S_b \quad \text{(mm/dry spell)}$$
Eq. 7.9

For $S_{start,i} > S_b$, transpiration is potential as long as the soil moisture content is greater than S_b. During that period the soil moisture content decreases linearly until the wet spell interrupts or until it reaches S_b at a time $t_{b,\,dry,i}$. Thus, starting the dry spell with $S_{start,dry} > S_b$, the soil moisture content will reach S_b at $t_{b,dry,i}$ unless the dry spell is interrupted. For $S_{start,\,dry} < S_b$, the moisture constrains transpiration from the start of the dry spell onwards, thus $t_{b,dry,i} - t_{dry,i} = 0$. Consequently:

$$t_{b,dry,i} - t_{dry,i} = \text{Max}\left(\frac{S_{start,dry,i}}{T_{pot}} - \gamma, 0\right) \qquad \text{(days)}$$
Eq. 7.10

$$T = T_{pot} \qquad \qquad \text{for } \text{Min}(t - t_{b,dry,i} + t_{dry,i}, \mu_{n_d})\ \text{(mm/day)}$$
Eq. 7.11

For $S_{start,dry,i} > S_b$, the total transpiration during a dry spell reads:

$$\int_{t_{dry,i}}^{t_{dry,i}+\mu_{n_{dry}}} T(t)\, dt = \left(t_{b,dry,i} - t_{dry,i}\right) * T_{pot} + S_b - S_b \exp\left(-\frac{1}{\gamma}\left(\mu_{n_{dry}} - t_{b,dry,i} + t_{dry,i}\right)\right)$$

$$\underset{S>S_b}{\longleftrightarrow} \qquad \underset{S<S_b}{\longleftrightarrow} \qquad \text{(mm/dry spell)}$$

$$= S_{start,dry,i} - S_b \exp\left(-\left(\frac{\mu_{n_{dry}}}{\gamma} - \frac{S_{start,dry,i}}{S_b}\right) - 1\right)$$
Eq. 7.12

The monthly rainfall P_m affects the average length of a dry spell.

Transpiration during a wet spell

During a wet spell, the soil moisture is continuously replenished by effective rainfall. Therefore, the change in soil moisture content has a 'source term'. For given soil conditions and monthly rainfall and for moisture constrained conditions:

$$\frac{dS}{dt} = -T + \mu_{P_{r,eff}} = -\frac{S}{\gamma} + \mu_{P_{r,eff}} \qquad \text{for } S < S_b \qquad \text{(mm/day)}$$
Eq. 7.13

For integration purposes, Eq. 7.13 is rewritten as

$$\frac{dS}{dt} = -\frac{S}{\gamma} + \frac{\mu_{P_{r,eff}} * \gamma}{\gamma} \qquad \text{for } S < S_b \qquad \text{(mm/day)}$$
Eq. 7.14

Eq. 7.14 has a standard solution that reads:

$$S(t) = \mu_{P_{r,eff}} * \gamma + \left(S_{start,wet,i} - \mu_{P_{r,eff}} * \gamma\right)\exp\left(-\frac{t - t_{wet,i}}{\gamma}\right)$$ Eq. 7.15

where

$S_{start,wet,i}$ is soil moisture content at start of the i-th wet spell   (mm)

$t_{wet,i}$ time at start of i-th wet spell   (days)

If $\mu_{P_{r,eff}} > T_{pot}$, the situation may occur that during the wet spell the boundary of S_b is surpassed. How long should a wet spell continue to have the soil moisture reach $S = S_b$? Transformation of Eq. 7.15 yields:

$$t_{b\uparrow,wet,i} - t_{wet,i} = -\gamma * \ln\left(\frac{S_b - \gamma * \mu_{P_{r,eff}}}{S_{start,wet,i} - \gamma * \mu_{P_{r,eff}}}\right)$$ Eq. 7.16

Be aware that realistic values of $t_{b\uparrow,wet,i} - t_{,wet,i}$ are positive. Because γ is also positive by definition (days), S_b is only surpassed if in Eq. 7.16 the ln-function is negative, which means that the ratio between brackets is between 0 and 1. This is by definition true because $\gamma * \mu_{P_{r,eff}} = S_b * \mu_{P_{r,eff}} / T_{pot} > S_b$.

As a consequence of Eq. 7.16, the transpiration total over the entire wet spell yields (mm/wet spell) for $S_{start,wet,i} < S_b$:

$$\int_{t_{wet,i}}^{t_{wet,i}+n_{wet}} T(t_{wet}) \, dt_{wet} = \mu_{P_{r,eff}} * \text{Min}(t_{b\uparrow,wet,i} - t_{wet,i}, \mu_{n_{wet}})$$ influx   Eq. 7.17

$$+ \left(S_{start,w} - \mu_{P_{r,eff}} * \gamma\right) * \left(\exp\left(-\frac{\text{Min}(t_{b\uparrow,wet,i} - t_{wet,i}, \mu_{n_{wet}})}{\gamma}\right) - 1\right)$$ change in storage

$$+ T_{pot,d}(\mu_{n_{wet}} - \text{Min}(t_{b\uparrow,wet,i} - t_{wet,i}, \mu_{n_{wet}}))$$ if S reaches S_b

For $S_{start,wet,i} > S_b$, moisture constrained conditions are only reached during the wet spell if mean potential transpiration is greater than the mean effective rainfall. Thus:

$$t_{b\downarrow,wet,i} - t_{wet,i} = \frac{S_{start,wet,i} - S_b}{T_{pot,d} - \mu_{P_{r,eff}}}$$ for $S_{start, wet,i} > S_b$ and $T_{pot} > \mu_{P_{r,eff}}$ otherwise $t_{b\downarrow,wet,i} \rightarrow \infty$ (days) Eq. 7.18

$$T = T_{pot,d}$$ for $\text{Min}(t_{b\downarrow,wet,i} - t_{wet,i}, \mu_{n_{wet}})$ (mm/day) Eq. 7.19

where $(t_{b\downarrow,wet,i} - t_{wet,i})$ is the time it takes from the start of the wet spell to reach the limit of moisture constrained transpiration, starting from a higher soil moisture content. If effective rainfall $\mu_{P_{r,eff}}$ is greater than $T_{pot,d}$, then the evaporation stays at $T_{pot,d}$ during the whole wet spell. As a result:

$$\int_{t_{wet,i}}^{t_{wet,i}+\mu_{n_{wet,i}}} T(t_{wet}) \, dt_{wet} = T_{pot,d} * \text{Min}(t_{b\downarrow,wet,i} - t_{wet,i}, \mu_{n_{wet}})$$ Eq. 7.20

$$+ \left(S_b - \mu_{P_{r,eff}} * \gamma\right) * \left(\exp\left(-\frac{\text{Max}(-t_{b\downarrow,wet,i} + t_{wet,i} + \mu_{n_{wet}}, 0)}{\gamma}\right) - 1\right)$$ for $S_{start, wet,i} > S_b$ (mm/wet spell)

7.5 Transpiration during a series of alternating dry and wet spells

In the preceding Sections analytical expressions for the total amount of transpiration during a dry and during a wet spell were derived. Therefore it is now possible to improve the monthly estimate of transpiration by a numerical calculation within the month.

In Figure 7.3 an example is given of a theoretical time series with certain initial conditions ($S_{start} < S_b$) and for certain static input parameters (S_b, $T_{pot,d}$, interception threshold D). The lengths of the numerical time steps vary with monthly rainfall and are alternating the mean length of a dry and of a wet spell. Under the given assumptions, for each spell an exact analytical solution is computed. The monthly transpiration is the integral of a compound function of the transpiration. Figure 7.3 illustrates how a change in monthly rainfall increases the length of the wet spells, decreases the length of the dry spells and increases the effective rainfall on rain-days. All these parameters are non-linear functions of the monthly rainfall and affect monthly transpiration. Potential transpiration has so far been considered constant at the end of the month, but will be studied in more detail in Chapter 10.

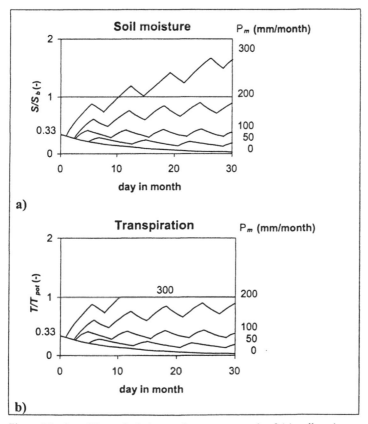

Figure 7.3 a,b   Theoretical time series over a month of (a) soil moisture and (b) transpiration for different monthly rainfall amounts P_m in Harare (mm/month), with an initial soil moisture of $0.33*S_b$ mm and an interception threshold D and potential transpiration T that are both 5 mm/day.

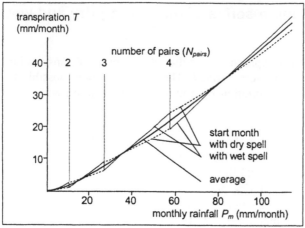

Figure 7.4   Illustration of the difference in monthly transpiration when initially starting with a wet or with a dry spell.

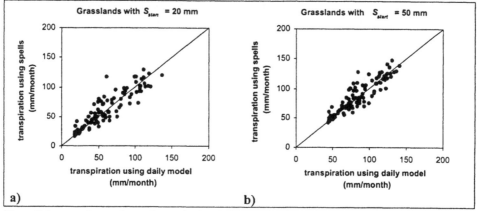

a)   b)

Figure 7.5   Comparison of transpiration during a month of a model with daily time steps and one with spells time steps, relying on monthly data. Use is made of rainfall and pan evaporation data for historical months from research station Grasslands (near Marondera). It is further assumed $S_{max} = 120$ mm, $S_b = 60$ mm, $D = 5$ mm/day. The outlier (60,123) is a month in which rainfall only occurred in the last part of the month. a) $S_{start} = 20$ mm, b) $S_{start} = 50$ mm.

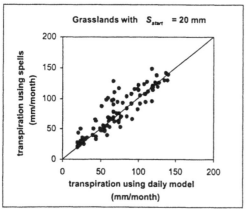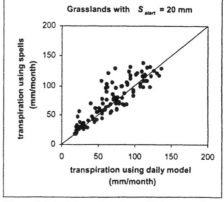

Figure 7.6   As Figure 7.5a, but with $S_b = 36$ mm instead of 60 mm, thus $S_b/S_{max} = 0.8$ instead of 0.5.

Figure 7.7 As Figure 7.5a but with interception threshold $D = 1$ mm/day instead of 5 mm/day.

Figure 7.4 illustrates the difference in monthly transpiration in the two different modes, which either start with a wet or with a dry spell. The function of each mode is discontinuous at the points where the expected number of cycles is an integer. To understand this, one should realise that the earlier water is available in the soil, the more the total transpiration during the month will be. Suppose the month has a number of pairs N_{pairs} that is just a little more than an integer. In that case, if the month starts with a wet spell, the first wet spell will be very short (see Eq. 7.3). This wet spell is followed by a dry spell of expected length. The total transpiration will then be close to that of a month which starts with a dry spell and which has a whole number of cycles. Therefore, for monthly rainfall for which the number of cycles is just more than an integer, transpiration is highest for the mode that starts with a dry spell, because that dry spell is very short. When N_{pairs} is just a little bit less than an integer, starting with a wet spell means beginning with a wet spell that is almost as long as the expected wet spell length. Thus, starting with a wet spell yields a higher transpiration than starting with a dry spell. This explains the discontinuity at whole numbers of N_{pairs}.

Figure 7.5 shows a comparison between a daily and a spells model for their estimations of monthly transpiration. Historical rainfall and pan evaporation data from the research station Grasslands were used. The difference between transpiration for the daily model and the spells model is less than 30 mm/month, except for one outlier, which had no rainfall in the first half of the month. There is no bias in the spells model. For the daily model the potential transpiration was assumed to be equal to the measured daily pan evaporation. For the spells model the monthly averaged daily pan evaporation was used. With use of constant potential transpiration in both daily and monthly models, the deviations between the two diminished considerably. Comparison of Figure 7.5 with Figures 7.6 and 7.7 shows that the performance of the spells model varies with $\gamma (= S_b/T_{pot})$ and with the interception threshold D.

Potential transpiration on dry days is generally higher than on rain-days, as will be explained in Chapter 10. However, the estimate of monthly transpiration, using the spells model, does not change if the monthly potential transpiration is attributed to the dry and wet spells with a different intensity.

7.6   Influence of variables on the spells model

The numerical computations with spells of expected length and expected density yielded monthly transpiration estimates that showed no bias compared with the daily model. The question is how the monthly results of the spells model relate to its input. The initial soil moisture condition $S_{start,m}$ is an output of the computation of the previous month. Thus it is time-variable and spatially-variable. Figure 7.8 illustrates that for lower amounts of rainfall, when the soil moisture reaches the constrained transpiration conditions for a part of the month, a change in initial soil moisture condition changes the monthly transpiration by an amount that is independent of the initial soil moisture. In the next Chapter it will be shown that this is true. For the very high monthly rainfall amounts, the potential transpiration becomes restrictive. For lower amounts of initial soil moisture, the transpiration approaches the potential monthly transpiration only at a higher monthly rainfall.

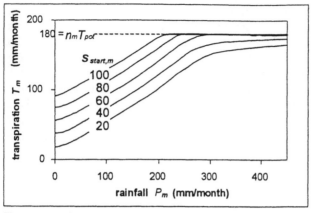

Figure 7.8   Transpiration per month T_m as a function of monthly rainfall P_m, for different values of initial soil moisture $S_{start,m}$. Same conditions as in Figure 7.5 and T_{pot} = 5 mm/day.

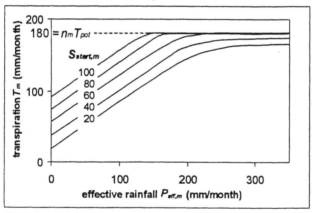

Figure 7.9   Monthly transpiration according to spells model for same conditions as in Figure 7.8, in this case plotted as a function of the monthly effective rainfall $P_{eff,m}$.

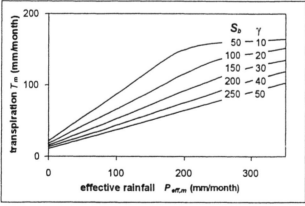

Figure 7.10   Transpiration per month T_m as a function of effective monthly rainfall $P_{eff,m}$, for different values of soil moisture S_b, with same T_{pot} = 5 mm/day and with S_{start} = 20 mm. Same conditions as in Figure 7.5.

Figure 7.9 shows that monthly transpiration T_m as a function of monthly effective rainfall $P_{eff,m}$ appears to be a linear relation. In the next Chapter it will be shown that this linearity is not exact, but that an approximation through a linear relation will make it possible to replace the spells model with a monthly model.

Figure 7.10 shows the results of the spells model for an initial soil moisture of 20 mm and different amounts of S_b, the soil moisture content at which transpiration becomes constrained. The potential transpiration T_{pot} is the same (5 mm/day). The value which determines the relation is $\gamma = S_b/T_{pot}$. The intercept with $P_{eff,m} = 0$ increases with decreasing values of γ. As potential transpiration is different for each month, γ is a time- and spatially-variable input parameter.

The length of dry and wet spell and the rainfall intensity, all as a function of the monthly rainfall P_m, are dependent on the spatially-varying coefficients of the power functions: p, q, r, v. The relations, however, prove insensitive to these parameters. This will be explained further in the next Chapter.

7.7 Conclusions

By dividing of the month in alternating time steps of the expected length of a wet spell and the expected length of a dry spell, an estimate for monthly transpiration can be obtained that incorporates the expected fluctuations in soil moisture availability within the month. Incoming fluxes of effective rainfall are assumed constant within a time step. From the start to the end of the time step the change in soil moisture, and thereby transpiration, can be computed analytically.

This Chapter showed that the spells approach for transpiration yielded transpiration estimations that do not have a bias when compared with results of a daily model. The model is useful as such, but plotting the monthly input and output values suggests that a simplified model can be developed for which it is not necessary to make a numerical computation within the month. In the next Chapter such a model is described and compared with established models for monthly transpiration.

8 Monthly Model for Transpiration

8.1 Monthly model

In the previous Chapter a model was derived to compute transpiration by dividing the month into time steps of expected spell lengths. Using this spells approach, in this Chapter a monthly model is developed.

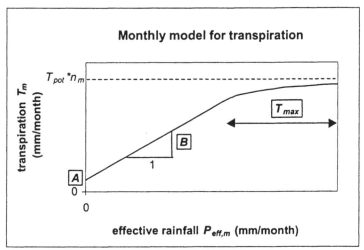

Figure 8.1   Monthly model as described in Eq. 8.1.

Figures 7.9 and 7.10 show that the relations between effective rainfall and transpiration per month are almost linear for the lower range of effective rainfall. With increasing monthly rainfall the slope gradually diminishes until at a certain point maximum transpiration is reached and the slope is zero. The equation for monthly transpiration as a function of monthly rainfall and initial soil moisture condition therefore has the following form:

$$T_m(P_m) = \text{Min}\left(A + B * (P_m - I_m), T_{max,m}\right) \qquad \text{(mm/month)} \qquad \text{Eq. 8.1}$$

where

A is the intercept with the $P_m = 0$ axis, which depends on the initial soil moisture $S_{start,m}$ and the potential transpiration $T_{pot,d}$.   (mm/month)

B slope of the linear part of the relation between monthly rainfall P_m and monthly transpiration T_m.   (-)

P_m monthly rainfall   (mm/month)

I_m monthly interception   (mm/month)

$T_{max,m}$ monthly transpiration restricted by potential transpiration and the time it takes to progress from moisture constrained to unconstrained transpiration   (mm/month)

What remains is to derive slope A, intercept B and the maximum monthly transpiration $T_{max,m}$ as a function of monthly rainfall P_m, maximum soil moisture content S_{max}, soil moisture content at which moisture constrained transpiration occurs S_b, initial soil moisture content at the start of the month $S_{start,m}$ and potential transpiration $T_{pot,d}$.

For notational convenience, the parameter γ (days) is transformed into a dimensionless parameter by dividing it with a time step of one month and with the number of days in a month n_m (days/month):

$$\gamma^o = \frac{\gamma}{n_m \Delta t_m} = \frac{S_b}{T_{pot,m} * \Delta t_m} \qquad (-) \qquad \text{Eq. 8.2}$$

where, as previously defined,

Δt_m is time in a month = 1   (month)

S_b available soil moisture content at the boundary between moisture constrained transpiration and potential transpiration   (mm)

$T_{pot,m}$ monthly potential transpiration, which is an input parameter   (mm/month)

The monthly potential transpiration is the sum of the values of daily potential transpiration for all days in the month:

$$T_{pot,m} = \sum_0^{n_m} T_{pot} \qquad \text{(mm/month)} \qquad \text{Eq. 8.3}$$

Intercept A

The intercept A with the $P_{eff,m} = 0$ axis (mm/month) can simply be derived by considering the total month as a dry spell (Eq. 7.13):

$$\boxed{A = \left(T_m | P_{eff,m} = 0\right) = \frac{S_{start,m}}{\Delta t_m} \left(1 - \exp\left(-\frac{n_m}{\gamma^o}\right)\right)} \qquad \text{for } S_{start,m} < S_b \qquad \text{Eq. 8.4}$$

(mm/month)

$$\boxed{\begin{aligned} A &= \left(T_m | P_{eff,m} = 0\right) \\ &= \frac{S_{start,m}}{\Delta t_m} - \frac{S_b}{\Delta t_m} \exp\left(-\frac{1 - \Delta t_{b,dry} / \Delta t_m}{\gamma^o}\right) \end{aligned}} \qquad \text{for } S_{start,m} > S_b \qquad \text{Eq. 8.5}$$

A value for $1/\gamma^o$ of around 2 is quite realistic. With 30 days in a month, γ is then 15 days. A change in the initial condition $S_{start,m}$ transmits with a ratio of $1 - e^{-2} = 0.86$ to the total transpiration over the month. A value of $1/\gamma^o > 2$, and thus an increase greater than 0.86, occurs when $\gamma < 15$. As an illustration, the realistic values $S_{max} = 120$ mm, $T_{pot} = 5$ mm/day and $S_b/S_{max} = 0.5$, yield $\gamma = 12$. Hence in practical cases A is almost equal to $S_{start,m}/\Delta t_m$; all soil moisture is transpired if it does not rain for the whole month.

For computer programming purposes Eqs. 8.4 and 8.5 can easily be combined to

$$A = \left(T_m | P_{eff,m} = 0 \right)$$

$$= \frac{S_{start,m}}{\Delta t_m} - \text{Min}\left(\frac{S_{start,m}}{S_b}, 1 \right)$$

for all S_{start} Eq. 8.6

$$* \frac{S_b}{\Delta t_m} * \left(1 - \exp\left(-\frac{1 - \text{Max}\left(\Delta t_{b,dry} / \Delta t_m, 0 \right)}{\gamma^\circ} \right) \right)$$

Slope B

The slope B is the rate (-) at which transpiration changes with effective precipitation. A change in effective precipitation affects the length of dry and wet spells and the average amount of rainfall on a rain-day. All these relate to effective monthly precipitation in a non-linear way, which is determined by the power relations between monthly rainfall and the conditional probabilities for rainfall occurrence p_{01} and p_{11}. Therefore it is surprising that the slope is almost constant for the major part of the range of effective rainfall.

Instead of disaggregating the month into dry and wet spells of the most probable length, it can be assumed that the month has constant rainfall. It appears that the subsequent analytical solution yields almost the same result for the monthly transpiration as the spells model. Within the month, the spells model is almost equal to the solution for constant effective rainfall at times when an equal number of wet and dry spells has passed since the start of the month. Figures 8.2 and 8.3 illustrate this. Thus, the ratio of the truncated pair of dry and wet spell to the full pair, which amounts frac(N_{pairs}) as in Eq. 7.2, is essential for the simplification to constant effective rainfall.

When effective rainfall is constant during the month, the analytical solution for slope B is simple. It means that the month is considered as one wet spell, similar to Eq. 7.15:

$$S_{end,m} = \frac{P_{eff,m}}{n_m}\gamma + \left(S_{start} - \frac{P_{eff,m}}{n_m}\gamma \right) * \exp\left(-\frac{n_m \Delta t_m}{\gamma} \right)$$

(mm)   Eq. 8.7

From the water balance it is clear that

$$T_m = P_{eff,m} + \frac{S_{start} - S_{end}}{\Delta t_m}$$

(mm/month)   Eq. 8.8

where $\Delta t_m = 1$ month. As a consequence, for a situation where $S < S_b$ during the whole month (mm/month):

$$T_m = P_{eff,m}\left(1 - \gamma^\circ + \gamma^\circ \exp\left(-\frac{1}{\gamma^\circ} \right) \right) + \frac{S_{start}}{\Delta t_m}\left(1 - \exp\left(-\frac{1}{\gamma^\circ} \right) \right)$$

Eq. 8.9

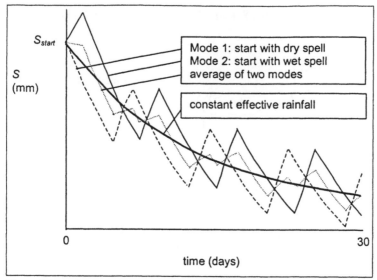

Figure 8.2   Theoretical time series over a month of soil moisture for the two different modes. Conditions are for Harare, with an initial soil moisture of 20 mm, a potential transpiration of T_{pot} = 5 mm/day , S_b = 60 mm and interception threshold D = 5 mm/day and P_m= 50 mm/month.

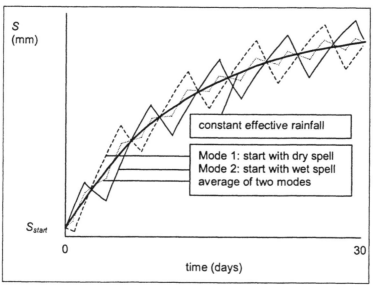

Figure 8.3   As Figure 8.2, but with monthly rainfall of 300 mm/month instead of 50 mm/month.

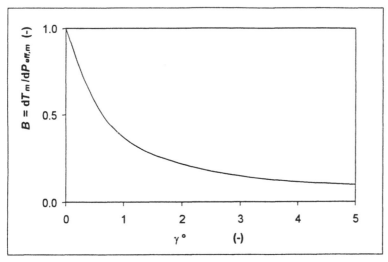

Figure 8.4   Relationship between γ° and slope B, the slope between effective rainfall and transpiration.

The second term is the intercept A, which is different when the month starts with $S_{start} > S_b$. The first term yields the solution for the slope B of the linear part of the relationship between effective rainfall and monthly transpiration. Thus is obtained:

$$B = 1 - \gamma^\circ + \gamma^\circ \exp\left(-\frac{1}{\gamma^\circ}\right)$$ (-)   Eq. 8.10

The relation is depicted in Figure 8.4. The slope B is independent of the amount of soil moisture at the start of the month S_{start}. This is also true for the exact solution, see Box 8-A. As described in the paragraph on intercept A, a realistic value of $1/\gamma^\circ$ is in the order of 2, or $\gamma/(n_m\Delta t_m) \approx 0.5$, which implies that $B \approx 0.57$.

For $\gamma^\circ \downarrow 0$ the slope B approaches 1, which means that every increase in effective rainfall is translated into transpiration.

Transpiration at very high monthly effective precipitation $T_{max,m}$

For $S_{start,m} > S_b$ and very high effective rainfall the maximum monthly transpiration is simply a full month with potential transpiration, because the soil does not suffer moisture constrained transpiration conditions.

$$T_{max,m} = T_{pot,d} * n_m$$ $S_{start,m} > S_b$ (mm/month)   Eq. 8.11

For $S_{start} < S_b$ it requires part of the month for the soil to reach moisture constrained transpiration conditions. Once the soil moisture is above S_b, transpiration will remain at the potential level during the month. This happens when:

$$\mu_{P_{eff,r}} * \mu_{n_w} > T_{pot} * (\mu_{n_w} + \mu_{n_d})$$ (mm/pair)   Eq. 8.12

Box 8-A   Exact solution of slope *B*

The exact solution for slope *B* as a function of *q, r, u, v, P_m, D* and γ can be computed using a mathematical computer programme such as MAPLE.

Figure 8A.1 presents an example of the slope as a function of the effective rainfall. As suspected, the slope is not constant and thus the relationship between effective monthly rainfall and transpiration is not linear. The slope as a function of effective rainfall is discontinuous where the number of pairs is an integer. However, the variation in the slope is small, even for low effective rainfall. In this example, the slope varies between 0.58 and 0.7. With high rainfall, the slope becomes an almost constant 0.635, which is between the local minima and maxima at lower monthly rainfall. With very high monthly rainfall, the slope increases again. However, in the example shown the transpiration is not limited to potential transpiration for $S > S_b$. For higher rainfall amounts, the slope *B* is therefore not relevant. At $P_{eff,m}$ = 260 mm/month the number of pairs N_p is back to 4 (see Figure 4.16).

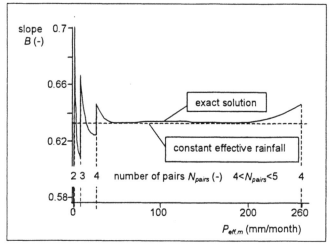

Figure 8A.1   Slope as a function of effective rainfall, for standard case Harare, S_b = 60 mm, T_{pot} = 5 mm/day, *D* = 5 mm/day.

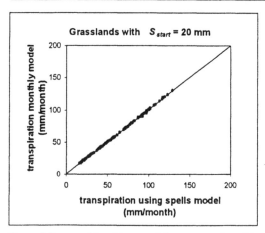

Figure 8.5   Monthly model compared with spells model for Grasslands (S_{max} = 120 mm, S_b = 60 mm, S_{start}=20 mm, *D* = 5 mm/day).

The assumption of constant rainfall during the month in this case also yields the most probable transpiration.

$$T_{max,m} = \frac{1}{\Delta t_m}\left(T_{pot} * (n_m \Delta t - \Delta t_{b\uparrow,wet}) - S_b + S_{start,m} + \mu_{P_{eff,r}} * \Delta t_{b\uparrow,w}\right)$$

$$S_{start} < S_b \text{ (mm/month)} \qquad \text{Eq. 8.13}$$

where $\Delta t_{b\uparrow,wet}$ is the time it takes to reach S_b.

Congruent with Eq.7.16 this $\Delta t_{b\uparrow,wet}$ can be expressed as:

$$\Delta t_{b\uparrow,wet} = -\gamma * \ln\left(\frac{S_b - \dfrac{\gamma}{n_m} * P_{eff,m}}{S_{start} - \dfrac{\gamma}{n_m} * P_{eff,m}}\right)$$

$$\begin{array}{c} S_{start} < S_b \\ \text{(mm/month)} \\ \text{Eq. 8.14} \end{array}$$

Thus the transpiration that is restricted by potential transpiration amounts:

$$T_{max,m} = \frac{1}{\Delta t_m}\left[\begin{array}{l} T_{pot} * n_m \Delta t - S_b + S_{start,m} \\ \\ -\left(\dfrac{P_{eff,m}}{n_m} - T_{pot}\right) * \gamma * \ln\left(\dfrac{S_b - \dfrac{\gamma}{n_m} * P_{eff,m}}{S_{start} - \dfrac{\gamma}{n_m} * P_{eff,m}}\right)\end{array}\right]$$

$$\begin{array}{c} S_{start} < S_b \\ \text{(mm/month)} \\ \text{Eq. 8.15} \end{array}$$

When programming the equation for monthly transpiration, one needs to be aware that for certain amounts of monthly rainfall $(S_b - \gamma/n_m * P_{eff,m})/(S_{start} - \gamma/n_m * P_{eff,m})$ becomes negative. Because a negative value does not have a natural logarithm, Eq. 8.15 will not yield a value. This only happens, however, in the linear part of Eq. 8.1, thus where $T_{max,m}$ does not apply.

Performance

Figure 8.5 illustrates the performance of the monthly model compared with the spells model ($R^2 = 0.999$, average relative error is 0.4%, maximum relative error is 3.6%, maximum absolute error is 3.1 mm/month). Here an interception threshold D of 5 mm/day is used, but the performance changes little with other realistic threshold values.

The results of the spells and the monthly models are almost the same, while the computations of the monthly model are far easier to make. In particular for distributed monthly models using GIS, the monthly equation is more practical.

8.2   Closing remarks

In the monthly model all parameters that depend on monthly disaggregation have been abandoned. Chapter 7 supports the derivation of the monthly model, but is of no use otherwise.

The simplicity of the final monthly solution is remarkable. The assumption of a constant flux of effective rainfall to the soil moisture is straightforward, but has not been applied up to now in hydrological models for water resources management. There is similarity to the evapoclimatonomy model, used in climate studies (Lettau, 1969; Lettau et al., 1979; Lettau and Hopkins, 1991; later adjusted by Nicholson, Kim et al., 1997; Nicholson, Lare et al., 1997). The basic equation of the evapoclimatonomy model reads:

$$\frac{dS}{dt} = P_m - E_{m1} - R_{m1} - \frac{S}{t^*} \qquad\qquad \text{mm/month} \qquad \text{Eq. 8.16}$$

where

E_{m1} is the immediate evaporation, roughly corresponding to the interception I_m (Nicholson, Lare et al., 1997).   (mm/month)

R_{m1} the immediate runoff   (mm/month)

t^* the average residence time of water in the soil, including groundwater (months). Thus, $t^* = \overline{S}/(\overline{E_{m2}} + \overline{R_{m2}})$, where $\overline{E_{m2}}$ is the long term average of the delayed evaporation (mm/month), roughly corresponding to transpiration T_m, and $\overline{R_{m2}}$ is the long term average of the delayed runoff, roughly corresponding to baseflow. In the evapoclimatonomy model both the delayed runoff and the delayed evaporation (transpiration) vary directly in proportion to the soil moisture content S. The residence time t^* shows similarities to the time scale γ in the model of this dissertation. However, the evapoclimatonomy model assumes constrained transpiration as soon as the soil moisture is at less than maximum. Thus, $S_{max} = S_b$. Also, groundwater recharge, eventually leading to delayed runoff, is in proportion to the soil moisture content, rather than occurring at saturation of S_{max}. The evapoclimatonomy model further assumes climatic stability, computing E_{m2} as a residual. Thus, no explicit equation for E_{m2}, i.e. transpiration, is derived.

In other monthly models the rate of transpiration during the month is not influenced by infiltration of effective rainfall:

- Thornthwaite & Mather (1955, 1957) presume that in months with rainfall at less than potential evaporation, all rain is evaporated. This method does not make a distinction between interception and transpiration and thus the evaporated rain is not necessarily only interception. Additionally, the soil moisture at the start of the month contributes to transpiration through a proportional relationship between soil moisture content and potential transpiration, which is the difference between potential evaporation and rainfall ($S_b = S_{max}$, $T_{pot} = (E_{pot,m} - P_{pot,m})/n_m$). Even with a very dry soil at the start of the month and only one rain-day with heavy rainfall, all rain is evaporated if the total rainfall is less than the monthly total potential evaporation. Alley (1984) uses the same approach, except that a fraction of the precipitation is immediately transformed into direct runoff.

- Palmer (1965, see Alley, 1984), Thomas (1981, see Alley, 1984) and Xu (1992) do not make a distinction between rainfall during the month and soil moisture available at the start of the month, therefore all rain is implicitly assumed to fall at the start of the month and a correction is made through calibration of parameters.

In models on smaller time scales the transpiration during the wet spell is normally also assumed at a potential rate (e.g. Eagleson, 1978).

In the next Chapter these and other monthly water balance models are discussed in more detail and their performance is compared with the monthly transpiration model suggested in this dissertation.

It needs to be stressed that the daily model on which the derived monthly transpiration model is based is a conceptual model, which means that physical process dynamics are represented by analytically-tractable solutions (definition by Troch et al., 1993). The transition from a daily to a monthly model made it necessary to further simplify. Average transpiration from an area with heterogeneity in soils and land cover is not well represented by a 'representative soil' (Kim, 1995). Remote Sensing offers a practical tool to determine spatially distributed 'soil' types. Optionally, areas can be clustered in 'hydrotopes' (Wolski, 1999).

9 Comparison with Thornthwaite-Mather and Other Monthly Models

9.1 Introduction

The great advantage of the modelling approach suggested in this dissertation is that it does not combine the evaporation processes interception and transpiration. Except for the evapoclimatonomy model, briefly described in Section 8.2, other monthly water balance models do not distinguish between the two. Therefore, a comparison with other models is restricted to total evaporation.

In the previous chapters the performance of the monthly model was compared with the performance of a daily model that was based on the same assumptions. The problem of comparison with other monthly models is that in general they are based on different assumptions. For example, all other monthly models use simplifications of the linear relationship between soil moisture availability and transpiration, Eq. 7.1. The performance criterion usually considers the reproduction of streamflow (e.g. Xu, 1992). This has the disadvantage that the groundwater module can compensate for deviations in the interception and soil moisture modules. The author is not aware of any publication in which a monthly model for evaporation is compared with a daily model.

Despite the difficulties mentioned above, in this chapter various monthly evaporation models are compared with the monthly model in this dissertation. Most attention is paid to the Thornthwaite-Mather model, as it is the most commonly used and because it also constrains evaporation when limited soil moisture is available.

The data from Grasslands experimental station, already shown in Figures 7.5 to 7.7, serve as input. For potential evaporation the pan evaporation data are used. Furthermore, in the basic case S_{max} = 120 mm, S_b = 60 mm, D = 5 mm/day and S_{start} = 20 mm.

9.2 Total evaporation according to Thornthwaite-Mather

Thornthwaite & Mather (1955, 1957) developed a model for evaporation that is still commonly used in current hydrological models at monthly time steps (e.g. Kwadijk & Van Deursen, 1999; Hoekstra, 1998; Vörösmatry & Moore, 1991; Thompson, 1992; Gleick, 1987).

Mather (1981, cited by Alley, 1984) adjusted the Thornthwaite-Mather equation to include information on possible overland flow during intense short-period precipitation events. Thomas (1981, as cited by Alley, 1984) illustrates how to adjust monthly precipitation to account for periods when much of the precipitation falls during the last part of the month. Both modifications require daily data, therefore they will not be examined further.

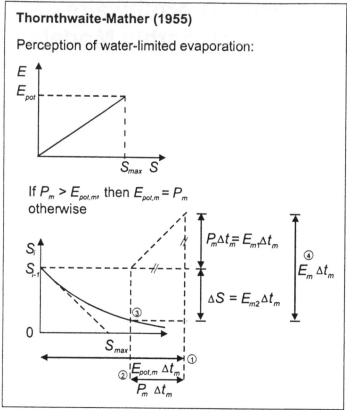

Thornthwaite-Mather (1955)

Perception of water-limited evaporation:

If $P_m > E_{pot,m}$, then $E_{pot,m} = P_m$
otherwise

Figure 9.1   Graphical description of Thornthwaite-Mather model. Procedure: ① Determine potential monthly evaporation $E_{pot,m}$. ② Subtract monthly rainfall P_m. ③ If the difference is more than zero, then soil moisture available at the end the month S_i is an exponential function of soil moisture at the start of the month S_{i-1}, dependent on maximum soil moisture in root zone S_{max}. ④ The actual evaporation is the sum of the depletion of the soil moisture and the rainfall.

The model assumes evaporation to be potential in months with more rainfall than the potential evaporation (Thornthwaite potential evaporation or an alternative). The excess amount of rain replenishes the soil moisture. If the maximum soil moisture content S_{max} is reached, the surplus causes surface runoff or recharge, similar to the model in this dissertation. For monthly rainfall lower than monthly potential evaporation, all rain is assumed to evaporate and furthermore the soil moisture content is depleted. Thus, within the Thornthwaite-Mather model two 'types' of evaporation can be distinguished: $E_m = E_{m1} + E_{m2}$. One part, here designated E_{m1}, is independent of the initial soil moisture and only depends on rainfall in the same month.

$$E_{m1} = \text{Min}\left(E_{pot,m}, P_m\right) \qquad\qquad \text{mm/month} \qquad \text{Eq. 9.1}$$

The difference between E_{m1} and the interception equation in this dissertation is that E_{m1} has an upper limit of monthly potential evaporation $E_{pot,m}$, whereas the interception presented in this dissertation has a daily upper limit of D. Hence in the Thornthwaite-Mather approach E_{m1} contains evaporation which in the method presented here is partly considered to be transpiration subject to the soil moisture balance equation.

In the Thornthwaite-Mather model E_{m2} is the part that depletes the soil moisture, which only occurs when $P_{i,m} < E_{pot,m}$. It is assumed that depletion of soil moisture during the month is proportional to the available soil moisture, following Penman (1948):

$$\frac{dS}{dt} = -\left(E_{pot,m} - P_m\right)\frac{1}{n_m}\frac{S}{S_{max}} \qquad \text{for } P_m < E_{pot,m} \quad \text{(mm/day)} \qquad \text{Eq. 9.2}$$

This assumption is a simplification of Eq. 7.1. Substitution of $T_{pot} = -(E_{pot,m} - P_m)/n_m$ and $S_b = S_{max}$ in the monthly model in this dissertation makes the assumption of constrained transpiration the same for both models. The simplification of Thornthwaite & Mather assumes that $S_b/S_{max} = 1$, while it is usually between 0.5 and 0.8 (see Section 7.2).

According to Thornthwaite & Mather, in months in which the rainfall P_m is less than the potential evaporation $E_{pot,m}$, the actual evaporation is $E_m = P_m - \Delta S/\Delta t_m$. As $E_{m1} = P_m$ for all months in which $P_m < E_{pot,m}$, it follows from Eq. 9.2 that:

$$E_{m2} = \frac{-\Delta S}{\Delta t_m} = \frac{S_{start}}{\Delta t_m}\left(1 - \exp\left(-\frac{\left(E_{pot,m} - P_m\right)\Delta t_m}{S_{max}}\right)\right) \qquad \text{(mm/month)} \qquad \text{Eq. 9.3}$$

$$\text{for } P_m < E_{pot,m}, \text{ otherwise } E_{m2} = 0$$

Figure 9.1 gives a graphical presentation of the procedure to determine the actual evaporation.

E_{m2} is similar to the transpiration equation in this dissertation in a situation where the total month is one dry spell, Eq. 7.9. The difference is that in the Thornthwaite-Mather model the potential 'transpiration' is reduced by evaporation that stems directly from the rainfall, i.e. E_{m1}. There is no distinction between wet-surface and dry-surface potential evaporation, thus E_{m1} consumes part of the potential evaporation. In this dissertation interception is assumed not to restrict potential transpiration, although the equation could very easily be adjusted to do so. A discussion on the more correct assumption follows in Chapter 10. In Figure 9.2a, b, the results of the basic case are shown for the Thornthwaite-Mather model and the model in this dissertation. For high amounts of evaporation the total amount of evaporation according to the Thornthwaite-Mather model in Figure 9.2a is restricted by the potential evaporation, while the daily model and the monthly model in Figure 9.2b have additional evaporation from interception.[29]

---

[29] The soil moisture storage at the end of month i is:

$$S_i = \min\left(\left(P_{i,m} - E_{pot,m}\right)\Delta t_m + S_{i-1}, S_{max}\right) \qquad \text{for } P_i > E_{pot}$$

$$S_i = S_{i-1}\exp\left(\frac{-\left(E_{pot,m} - P_{i,m}\right)\Delta t_m}{S_{max}}\right) \qquad \text{for } P_i < E_{pot}$$

If the annual potential evaporation is greater than the annual rainfall and if the method is used to determine evaporation during an average year, the Thornthwaite-Mather method requires iteration of the start value for S_{i-1}. Thornthwaite & Mather introduce a parameter that is called accumulated potential water loss ($APWL$). Although this parameter is still used (e.g. Hoekstra, 1998), the representation above is more transparent and is in line with Alley (1984). The author has introduced the parameter Δt_m to represent the units correctly.

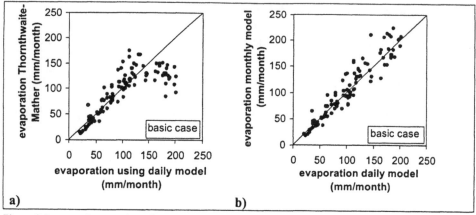

Figure 9.2   Scatter plot of evaporation of monthly models to daily model for basic case. a) Thornthwaite-Mather model b) this dissertation. Parameters basic case: S_{max} = 120, S_b = 60 mm, D = 5 mm/day and S_{start} = 20 mm. In the daily model, potential transpiration and potential interception are complementary. In the Thornthwaite-Mather model there is no distinction between interception and transpiration, therefore the evaporation is limited by the potential evaporation, which is equal to the potential transpiration in the daily model. This explains why the Thornthwaite-Mather estimate is considerably lower than the daily model.

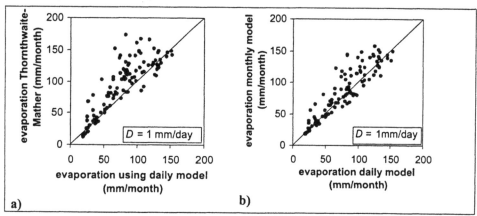

Figure 9.3   As Figure 9.2, but the interception threshold D has been changed from D = 5 mm/day to D = 1 mm/day.

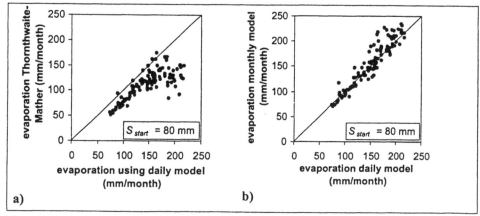

Figure 9.4   As Figure 9.2, but S_{start} = 80 mm instead of 20 mm.

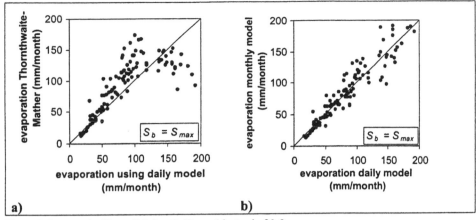

Figure 9.5   As Figure 9.2, but S_b/S_{max} = 1 instead of 0.5.

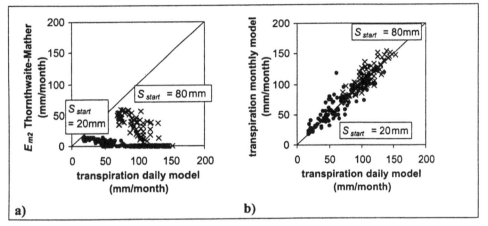

Figure 9.6   Scatter plot of transpiration in daily model with a) E_{m2} of Thornthwaite-Mather model b) monthly model in this dissertation. Two different starting values for soil moisture are depicted: • S_{start} = 20 mm, × S_{start} = 80 mm.

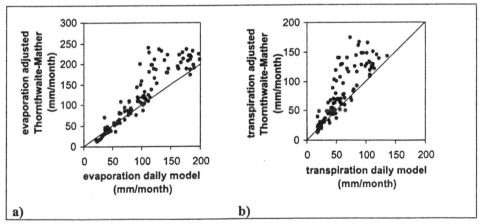

Figure 9.7   Scatter plot of daily model to adjusted Thornthwaite-Mather model for a) evaporation, b) transpiration.

The Thornthwaite-Mather model does not have a parameter for the interception threshold D. For the basic case a value of $D = 5$ mm/day was chosen. This value gives the fewest deviations between the daily model and the Thornthwaite-Mather model. For comparison, Figure 9.3a shows the results of $D = 1$ mm/day. However, the choice of $D = 5$ mm/day is arbitrary, because for a different start value of the moisture content, a different threshold value D would have been calibrated. Figure 9.4a,b shows that if the starting soil moisture content S_{start} is 80 mm instead of 20 mm, the Thornthwaite-Mather model is more biased than the model in this dissertation. Because one of the underlying assumptions of the Thornthwaite-Mather model is that $S_b = S_{max}$, the performance is tested under these non-realistic conditions. Comparison of Figure 9.2a with Figure 9.5a shows that the performance of the Thornthwaite-Mather model is worse when $S_b = S_{max}$ than when $S_b = 0.5*S_{max}$. Comparison of a) and b) in Figures 9.2 – 9.5 shows that in all cases the model in this dissertation performs better.

9.3 Transpiration derived from the Thornthwaite-Mather model

In the Thornthwaite-Mather model, E_{m2} is the part of evaporation that depends on the soil moisture status at the start of the month. The Thornthwaite-Mather model is implicitly based on the assumption that interception (including soil evaporation) is as great as the monthly precipitation if $P < E_{pot,m}$. If potential evaporation is not satisfied, additional evaporation comes from transpiration of the soil moisture. However, Figure 9.6a shows that E_{m2} is much less than transpiration in the daily model. This is logical, because 1) there is no replenishment of soil moisture during the month, 2) the potential transpiration is less, because precipitation is first abstracted, 3) in the monthly model transpiration is potential when $S > S_b$, while Thornthwaite-Mather assumes a reduction of transpiration for all $S < S_{max}$. It is interesting to note that for a $S_{start} = 80$ mm in comparison to a $S_{start} = 20$ mm, total evaporation in the Thornthwaite-Mather model is less than in the daily model (Figure 9.4a), while E_{m2} is more similar to the transpiration of the monthly model.

In Chapter 6, a separate equation for interception has been defined. It would thus be most logical to use this I_m and adjust the Thornthwaite-Mather equation in order to move closer to the assumptions in this dissertation. If it is assumed that the I_m, which is defined in Chapter 7, is the interception and that this interception does not limit transpiration, the Thornthwaite-Mather model (Eq. 9.3) can be adjusted to find an estimate for transpiration:

$$T_m = P_m - I_m - \frac{\Delta S}{\Delta t_m}$$

$$= P_m - I_m - \frac{S_{start}}{\Delta t_m}\left(1 - \exp\left(-\frac{\left(E_{pot,m} - P_m + I_m\right)\Delta t_m}{S_{max}}\right)\right)$$

(mm/month)
Eq. 9.4

Comparison of Figure 9.7a with Figure 9.2b shows that high amounts of evaporation are now overestimated instead of underestimated. Comparison of Figure 9.7b with Figure 9.6a shows that the estimate for transpiration improved considerably, but is too high. It can be seen from Figure 9.6b that the model in this dissertation performs better.

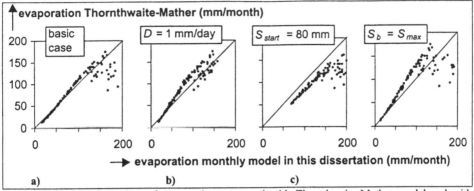

Figure 9.8   Scatter plots of evaporation computed with Thornthwaite-Mather model and with monthly model in this dissertation. a) basic case, b) $D =1$ mm/day instead of 5, c) S_{start}= 80 mm instead of 20, d) S_b=S_{max}=120 mm instead of S_b=60 and S_{max}=120 mm.

9.4   Linearity between Thornthwaite-Mather and model in this dissertation

For the smaller amounts of rainfall, in the linear part of Eq. 8.1, the total evaporation of the Thornthwaite-Mather model and the model in this dissertation relate quasi-linearly (in the statistical sense). The scatter plots of Figure 9.8 demonstrate this.

It should be realised that $1/\gamma$ (-) in the monthly model in this dissertation is equal to $T_{pot,m}\Delta t_m/S_b$ (-). Substituting this in Eq. 8.9 and adding interception yields the total evaporation in the model in this dissertation:

$$E_{Gm} = B*P_m + (1-B)*I_m + \frac{S_{start}}{\Delta t_m}\left(1-\exp\left(-\frac{T_{pot,m}\Delta t_m}{S_b}\right)\right)$$

Eq. 9.5

The Thornthwaite-Mather model can be rewritten as:

$$E_{TMm} = P_m + \frac{S_{start}}{\Delta t_m}\left(1-\exp\left(\frac{P_m\Delta t_m}{S_{max}}\right)\exp\left(-\frac{E_{pot,m}\Delta t_m}{S_{max}}\right)\right)$$

Eq. 9.6

Combining Eqs. 9.5 and 9.6 yields:

$$E_{Gm} = B*E_{Tm} + \left\{B*S_{start}-S_{start}\exp\left(-\frac{E_{pot,m}}{S_b}\right)+1\right\}$$
$$+\left\{(1-B)I_m + B*S_{start}\exp\left(-\frac{E_{pot,m}+P_m}{S_{max}}\right)\right\}$$

Eq. 9.7

As mentioned before, the Thornthwaite-Mather model uses for both $T_{pot,m}$ and $E_{pot,m}$ the same potential evaporation, which will usually be in the same order as pan evaporation data. For Grasslands monthly pan evaporation data ranged from 86 to 210 mm/month, while the monthly precipitation ranged from 0 to 460 mm/month.

Although the linearity between E_{Gm} and E_{TMm} is only valid for $P_m < 250$ mm/month, the variance in pan evaporation is far less than that in precipitation. Therefore $E_{pot,m}$ is almost constant and the first bracketed term can be regarded as the intercept of the relationship between between E_{Gm} and E_{TMm}. It is clear that B is part of the slope. The second bracketed term is in almost direct proportion to E_{Gm}. This explains the quasi-linearity.

9.5   Other models

Pitman

The Pitman rainfall-runoff model (Pitman, 1973) has been applied widely in Southern Africa. It now forms the hydrological module of the WRSM90 model (Hughes, 1997), which is a surface water resources planning model incorporating river networks and reservoir operation. Irrigation demands do not follow directly from the rainfall, but are input time series. Modelled evaporation thus only influences the output of streamflow.

As shown in Chapter 6, the Pitman model accounts for monthly interception separately. Surface runoff relates to rainfall with an S-curve function. The surplus of rainfall enters a stock that represents the sum of groundwater (G) and soil moisture (S). Both transpiration and runoff (R) relate to the combined stock of underground water. The baseflow runoff is a power function of the underground water stock, with the power more than 1. The baseflow does not reach the catchment outlet in the same time step, but is lagged with the use of the Muskingum procedure. If the underground water stock is less than a certain value, transpiration is assumed to be zero. The relation of transpiration with the soil moisture is as depicted in Figure 9.9.

Pitman's mathematical description can be simplified to the following equation, which is more analogous with the equations in this dissertation:

$$T_m = \text{Max}\left((S+G)_{start} - R, 0\right)$$
$$* \frac{T_{pot,m}}{(S+G)_{max} - R} \qquad \text{(mm/month)} \qquad \text{Eq. 9.8}$$

where
$(S+G)_{start}$ is the sum of groundwater and soil moisture at the start of the month.(mm)
$(S+G)_{max}$ the maximum of the sum of groundwater and soil moisture   (mm)
R monthly runoff   (mm/month)

$$R = \rho * (S+G)_0$$
$$= \rho * (S+G)_{max} * \frac{T_{pot,m}}{T_{pot,max,m}} \qquad \text{(mm/month)} \qquad \text{Eq. 9.9}$$

where
ρ is the portion of the total soil moisture and groundwater that runs off in the same month   (1/month)
$T_{pot,max,m}$ the maximum monthly potential transpiration in the year   (mm/month)

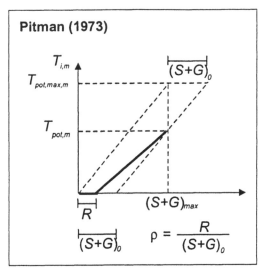

Figure 9.9   Graphical description of monthly transpiration model by Pitman (1973).

Pitman (1973) maintains ρ fixed for different rainfall regions in South Africa. Calibration for areas with summer rainfall that are humid and semi-arid yielded a value of $\rho = 0.5$ (1/month). The arid and the non-arid catchments with no rainy season or a winter rainy season had a value of $\rho = 0$. However, in applying the Pitman model, Hughes (1997) lets ρ depend on the vegetation cover with a seasonal change, which actually makes it easier to directly calibrate R and abandon the parameter ρ.

Because of the combining of soil moisture and groundwater, it is difficult to compare the Pitman model with the model in this dissertation. The Pitman model does not account for a change in the soil moisture, and thus the transpiration rate, within the time step. Instead, the model uses a numerical computation within the month. For South Africa, Pitman split the month into four time steps. A larger number of time steps barely changed the estimate of transpiration. The instantaneous soil moisture content at the start of the time step is supposed to determine the monthly transpiration.

Palmer

Palmer (1965) used two soil layers. The upper and lower layers have available moisture capacities, $S_{a,max}$ and $S_{b,max}$ respectively. Moisture cannot be removed from (recharged to) the lower layer until all moisture has been removed from (replenished in) the upper layer. All rainfall at less than the potential evaporation is assumed to evaporate in the same month. Additionally, for the surplus of potential evaporation, evaporation from the upper layer (E_a) takes place at the potential rate $E_{pot,m}$. Evaporation from the lower layer (E_b) occurs when the potential evaporation cannot be met by the storage in the upper layer.

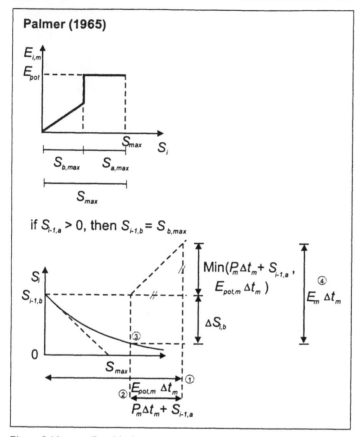

Palmer (1965)

if $S_{i-1,a} > 0$, then $S_{i-1,b} = S_{b,max}$

Figure 9.10 Graphical description model by Palmer (1965). Procedure: ① Determine potential monthly evaporation $E_{pot,m}$. ② Subtract monthly rainfall $P_m.\Delta t_m$ + soil moisture in upper soil layer $S_{i-1,a}$ ③ If the difference is more than zero, then soil moisture in the lower layer is available at the end of the month. $S_{i,b}$ is an exponential function of soil moisture at the start of the month $S_{i-1,b}$, dependent on maximum soil moisture in root zone S_{max}. ④ The actual evaporation is the sum of the depletion of the soil moisture in the upper and lower layers and the rainfall.

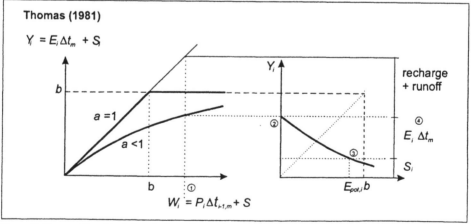

Thomas (1981)

$Y_i = E_i \Delta t_m + S_i$

$W_i = P_i \Delta t_{i-1,m} + S$

Figure 9.11 Graphical description of monthly evaporation model by Thomas (1981).

Thus:

$$E_m = \text{Min}\left(E_a + E_b + P_m, E_{pot,m}\right)$$

Eq. 9.10

$$E_a = \text{Min}\left(S_{a,start}, E_{pot,m} - P_m\right) \qquad \text{for } P_m < E_{pot,m}$$

Eq. 9.11

$$E_b = \left(\frac{S_{b,start}}{S_{max}}\right)\left(E_{pot,m} - P_m - E_a\right) \qquad \text{for } P_m + E_a < E_{pot,m}$$

Eq. 9.12

The principle is depicted in Figure 9.10. Recharge to zone b only takes place when a is full. Runoff is assumed to occur if and only if soil moisture storage in both layers has reached its capacity. A daily hydrological model by Wilby et al. (1994) is based on the same principles, although Wilby refers directly to Penman (1948). Calder et al. (1983) use a three-layer model of the soil moisture, where if there is moisture in the upper reservoir, evaporation is potential. If not, but there is water in the second reservoir, evaporation is half potential. If there is only water in the third reservoir, evaporation is quarter of potential. Moisture exchange between reservoirs is through cascading.

Thomas

Thomas (1981, publication not available, described by Alley, 1984) presented a monthly evaporation model that puts an upper limit on the total of evaporation and soil moisture storage, rather than on soil moisture storage and on evaporation separately. Thomas called his model the 'abcd' model. The method served for an improved water resources assessment of the USA and thus should suit different climatic, land cover and geological circumstances.

The sum of precipitation and soil moisture storage W_i is allocated between actual transpiration and build-up of soil moisture storage, with the assumption that the rate of loss of soil moisture to transpiration is proportional to the soil moisture storage, in the way Thornthwaite & Mather did. The equations in Thomas's model can be transformed into a direct expression for transpiration congruent with the descriptions above (after Alley, 1984; the author of this dissertation added Δt_m to obtain consistency in units):

$$E_{i,m} = \frac{Y_i}{\Delta t_m}\left(1 - \exp\left(\frac{-E_{pot,m}}{b}\right)\right)$$

Eq. 9.13

where

$$Y_i(W_i) = S_i + E_{i,m}\Delta t_m = \left(\frac{W_i + b}{2a}\right) - \left(\left(\frac{W_i + b}{2a}\right)^2 - \frac{W_i b}{a}\right)^{0.5}$$

Eq. 9.14

and

$$W_i = P_{i,m} * \Delta t_m + S_{i-1}$$

Eq. 9.15

The parameter b (mm/month) is an upper limit of the total of monthly transpiration and soil moisture storage. This implies that $b = S_{max}/\Delta t_m + E_{pot,m}$. The maximum value for Y_i is also $S_{max} + E_{pot,m}\Delta t_m$. Consequently, the actual evaporation for $P_m \rightarrow \infty$ does not reach $E_{pot,m}$.

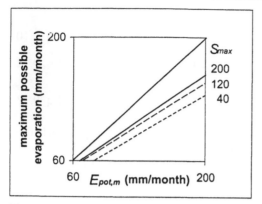

Figure 9.12   Maximum possible evaporation in Thomas's model, given certain conditions of maximum soil moisture content S_{max} and monthly potential transpiration $E_{pot,m}$.

The maximum actual evaporation is:

$$E_{max} = \left(\frac{S_{max}}{E_{pot,m} \Delta t_m} + 1 \right) * \left(1 - \exp\left(- \frac{1}{\dfrac{S_{max}}{E_{pot,m} \Delta t_m} + 1} \right) \right)$$ Eq. 9.16

The difference between E_{max} and E_{pot} can be as great as 35%, see Figure 9.12. Alley (1984) realised that evaporation could not reach potential evaporation, but did not quantify the minimum difference. Normally, the physical connotation of b is disregarded and b is determined by calibration with streamflow. With Eq. 9.16 in mind, it is not surprising that both Alley (1984) and Xu (1992) calibrated very high values for b (Alley: $356 < b < 1270$ mm, for catchments in New Jersey, Xu: $260 < b < 1900$ mm, for catchments in China and Belgium). It is unrealistic to suggest that for these values $b = S_{max}/\Delta t_m + E_{pot,m}$.

The parameter a ($0 < a < 1$) reflects 'the propensity of runoff to occur before the soil is fully saturated' (after Alley, 1984; reference to Thomas et al., 1983). If a is 1, runoff does not occur before the soil is fully saturated. Both Alley and Xu found values close to 1 (Alley: $0.975 < a < 0.999$; Xu: $0.96 < a < 0.999$). In the model in this dissertation, $a = 1$. However, because Alley and Xu have values for b that are unrealistic, values for a are bound also to lose their physical connotation. In Thomas's model, the difference W_i - Y_i is the sum of direct runoff and groundwater recharge.

Alley

Alley (1984) proposes two variations on the Thornthwaite-Mather model. The first is to turn a proportion of the rainfall into direct runoff. The model in this dissertation can easily be adjusted in a similar way, if there is a reason to do so. The second variaton is to use a combination of the model by Thornthwaite-Mather and the model by Thomas. Given a certain soil moisture amount at the start of the month, the solution is exactly as in the Thornthwaite-Mather equation. The difference is in the build-up of the soil moisture. First evaporation is computed with the use of the Thornthwaite-Mather equation. The difference comes with the computation of soil moisture at the end of the month. Analogous to Eq. 9.14, for Alley's model available moisture is defined as:

$$W_i' = (P_{i,m} - E_{i,m})\Delta t_m + S_{i-1}$$ Eq. 9.17

Consequently, the soil moisture at the end of the month amounts to:

$$S_i(W_i') = \left(\frac{W_i' + S_{max}}{2a}\right) - \left(\left(\frac{W_i' + S_{max}}{2a}\right)^2 - \frac{W_i' * S_{max}}{a}\right)^{0.5}$$ Eq. 9.18

The values Alley found were very different for the different catchments, covering the full range of possible a's from 0 to 1 (Alley: $0.014 < a < 0.99$). The procedure can easily be applied to the model in this dissertation, if necessary.

Xu

Xu (1992; see also Vandewiele et al., 1992) developed two new monthly evaporation models. He calibrated the models for different catchments in Belgium and China and found his own superior to those of others, using the reproduction of streamflow as the performance criterion.

Similar to Thomas (1981), in Xu's models evaporation is a function of the 'available moisture' $W_i = P_{i,m}\Delta t_m + S_{i-1}$.

The actual evaporation is in the first Xu model:

$$E_i(W_i) = \text{Min}\left(E_{pot,m}\left(1 - a^{W_i/E_{pot,m}}\right), W_i\right)$$ Eq. 9.19

and in the second Xu model:

$$E_i(W_i) = \text{Min}\left(W_i\left(1 - \exp(-aE_{pot,m})\right), E_{pot,m}\right)$$ Eq. 9.20

Both models are purely empirical. The parameter a does not have a physical meaning in either of the models. However, when a is regarded as $1/S_{max}$, then Eq. 9.20 (Xu 2) shows great similarities to both the Thornthwaite-Mather model and the model in this dissertation. This is discussed further below.

Makhlouf & Michel

Makhlouf & Michel (1994) developed a monthly water balance model, using data from 91 catchments in France. They evaluated the performance using the Nash criterion for streamflow reproduction. Makhlouf & Michel concluded that their model is better than the Thornthwaite-Mather model, Thomas's 'abcd' model and Xu's model. Their monthly evaporation is independent of the initial soil moisture content at the start of the month:

$$E_m = E_{pot,m} - \frac{P_m * E_{pot,m}}{\left(P_m^{0.5} + E_{pot,m}^{0.5}\right)^2}$$ Eq. 9.21

Eq. 9.21 implies that with a particular potential monthly evaporation any increase in rainfall will lead to a decrease in actual evaporation. This is highly unrealistic and therefore the model is not further discussed here.

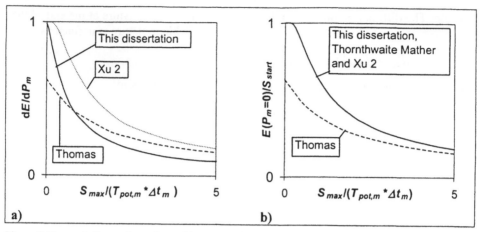

Figure 9.13   a) Comparison of monthly evaporation in different models, when monthly rainfall = 0. b) Comparison of change in monthly evaporation as a result of change in monthly rainfall. Both comparisons are made for different values of $S_{max}/(T_{pot,m}*\Delta t_m)$. For the model in this dissertation it is assumed that $S_b=S_{max}$ and there is no interception ($D = 0$). For Thomas's model, $a = 1$, so no recharge or runoff when S_{max} is not reached, as in the model in this dissertation. Additionally $b = S_{max} + E_{pot.m}$. For Xu's second model it is assumed that $a = 1/S_{max}$, similar to the other models.

Comparison of slopes and intercepts

In the model in this dissertation, monthly transpiration relates linearly to monthly effective rainfall, for low rainfall amounts. In this section how different models react when the monthly rainfall P_m is zero is described. In addition, for the models that relate linearly to (effective) rainfall, the slopes are compared.

To arrive at similar underlying assumptions to those in most models, the interception threshold D is set to zero. Consequently, potential evaporation is equal to potential transpiration ($T_{pot,m}= E_{pot,m}$). For the same reason, evaporation is supposed to be constrained as soon as soil moisture is not at maximum ($S_b = S_{max}$). Additionally, in Xu's second model the parameter a is assumed to be equal to $1/S_{max}$.

The amount of transpiration that occurs when monthly rainfall is zero (A) is expressed as a function of the initial soil moisture content S_{start} and of the dimensionless time scale γ° (Eqs. 8.4 and 8.5). This is equivalent to $S_{max}/(E_{pot,m}\Delta t_m)$ in the models that do not make a distinction between E_{pot} and T_{pot} or between S_b and S_{max}. The Thornthwaite-Mather model, the second model by Xu and the model in this dissertation have the same evaporation in months when $P_m = 0$. Analogous to Eq. 8.5, when rainfall is zero evaporation amounts to:

$$\frac{E(P_m = 0)}{S_{start}} = 1 - \exp\left(-\frac{1}{\gamma^\circ}\right)$$

Eq. 9.22

For Thomas's model the equation is slightly different:

$$\frac{E(P_m = 0)}{S_{start}} = 1 - \exp\left(-\frac{1}{\gamma^\circ + 1}\right)$$

Eq. 9.23

The difference is shown in Figure 9.13b.

Evaporation at low rainfall amounts is considerably less in Thomas's model, even when $S_b = S_{max}$ and interception is disregarded. Thomas's model and Xu's second model also show a linear relationship between rainfall and evaporation. However, where the slope B of the transpiration model is

$$\frac{dE}{dP_m} = B = 1 - \gamma^\circ + \gamma^\circ \exp\left(-\frac{1}{\gamma^\circ}\right) \qquad \text{Eq. 9.24}$$

(see Eq. 8.8). For Thomas's model it reads:

$$\frac{dE}{dP_m} = 1 - \exp\left(-\frac{1}{\gamma^\circ + 1}\right) \qquad \text{Eq. 9.25}$$

Thus in Thomas's model the slope is the same as the intercept. Eq. 9.25 shows that Thomas's low intercept is not compensated by a steeper slope, therefore for any value of P_m the estimate by Thomas is lower than that by the model in this dissertation, see Figure 9.13a. For cases when a in Thomas's model is chosen to be less than 1, the difference will be even greater.

From Xu's second model can be derived:

$$\frac{dE}{dP_m} = 1 - \exp\left(-\frac{1}{\gamma^\circ}\right) \qquad \text{Eq. 9.26}$$

The slope is steeper than in this dissertation, see Figure 9.13a. Xu's models are purely empirical. A value of a may be calibrated that is higher than $1/S_{max}$, in a way that makes the slope similar to that in this dissertation. Alternatively, the steeper slope can compensate for the neglect of interception.

9.6 Conclusions

The most important conclusions that follow from the preceding Sections are:
- Monthly evaporation models do not make a distinction between transpiration and interception, with the exception of Pitman (1973).
- Many monthly evaporation models assume that as long as precipitation is less than potential evaporation it is completely evaporated in the same month (Thornthwaite & Mather, 1955; Palmer, 1965). Other monthly evaporation models do not distinguish between moisture available at the start of the month and the total rainfall during the month as sources for evaporation (Thomas, 1981; Xu, 1992).
- The model in this dissertation is faithful to the general insight that transpiration is not moisture constrained as long as the soil moisture is above a certain value (S_b). Only Palmer (1965) also accounts for unrestricted transpiration as long as soil moisture is above a certain threshold.
- All other monthly evaporation models disregard the influx of effective rain into the soil moisture during the month. The simple analytical solution that follows from the assumption of constant effective rainfall has not been employed before. Pitman's perception of evaporation (1973) is most similar to the model in this dissertation. Firstly, Pitman's model distinguishes between interception and transpiration. His empirical model yields similar results to the interception model in this dissertation (see Chapter 6). Secondly, through the numerical computations within the month, the assumption of uniform effective rainfall is approached. However, because groundwater and soil moisture are combined in Pitman's

model, the relationship between 'underground water' and transpiration is different from that between soil moisture and transpiration, which is used in this dissertation.

- The model presented in this dissertation is preferred over the other models because it performs best when compared with aggregated daily models.

10 Estimating Potential Transpiration

10.1 Introduction

In this Chapter it is shown how potential transpiration on dry days and on rain-days differs considerably. It is shown that by using this knowledge, through a combination of monthly pan evaporation data and monthly rainfall data, determining average potential transpiration within the month is improved. The expected monthly average of potential daily transpiration serves as input for the monthly transpiration model, which was described in Chapter 8.

In the Chapters on interception and transpiration, the performance of monthly models has been compared with daily models for different assumed values for the daily interception threshold (D) and the potential transpiration (T_{pot}). It is here investigated how the potential transpiration on rain-days is limited by actual interception, which consumes part of the available energy. Moreover, rain-days and dry days not only differ in the occurrence of rain, but also in other meteorological variables. The influence of these variables is determined as well.

In lumped deterministic hydrological models, whether monthly or daily, a ratio is usually calibrated that links potential evaporation for a certain land cover type to a reference potential evaporation. The calibrated ratios for different landcover types depend not only on the way in which this reference potential evaporation is determined (pan, Penman, Penman-Monteith, Thornthwaite, Makkink), but also on the model's relationship between potential evaporation and actual evaporation. Through this calibration the water balance is closed. In the most advanced case, In distributed deterministic models potential interception and transpiration are made a function of vegetation cover, soil type, soil depth and meteorological conditions. These functions also include parameters that are calibrated in order to close the water balance. Thus, potential transpiration is only indirectly verified in hydrological models.

It is beyond the scope of this dissertation to solve the difficulty of determining correct daily values for potential interception and potential transpiration. However, it is realised that the accuracy of such estimates determines the success of the monthly models proposed, as it does for any monthly or daily model.

In this Chapter, data from the meteorological station Grasslands are used as an illustration. Twenty years of daily rainfall and pan evaporation data are available for this research station (1978-1996), data which were also used in Chapters 7 and 8, where the potential transpiration was assumed to be equal to pan evaporation. Additionally, use is made of a database that contains daily records of all meteorological variables (rainfall, pan evaporation, average humidity, insolation, sunshine hours, maximum and minimum temperature) from seventeen meteorological stations throughout Zimbabwe for the years 1995-1998.

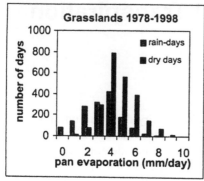

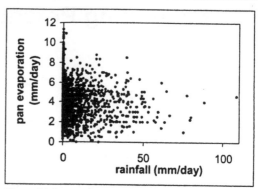

Figure 10.1   Histogram of the number of days within a class of daily pan evaporation, for rain-days and dry days separately. Data from Grasslands 1978-1998, rainy seasons Nov-April.

Figure 10.2   Scatter diagram of daily rainfall against pan evaporation for Grassslands. The scatter diagram shows no significant correlation.

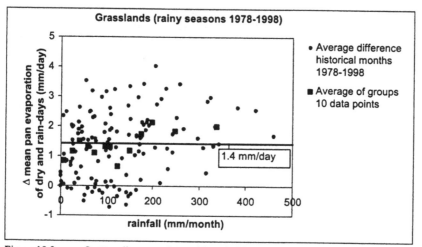

Figure 10.3   Scatter diagram of monthly rainfall against the difference of the mean pan evaporation of dry days and that of rain-days within the month, for Grasslands. The symbol + shows the averages of 13 consecutive groups of 10 data points.

10.2 Differences in pan evaporation between dry and rain-days

Figure 10.1 shows the histograms of pan evaporation on dry and rain-days in the wet season for the Grasslands station (1995-1998, November -April). In Zimbabwe, American Weather Bureau Class A pans are used as standard. All pans are covered by chicken wire to prevent animals drinking from the pan. The chicken wire reduces the pan evaporation by about 10% (Torrance, 1981).

Pan evaporation on rain-days is on average less than on dry days. For rain-days as such, Figure 10.2 shows that there is no significant relationship between daily rainfall and daily pan evaporation. Also, the difference between the mean evaporation on dry and rain-days within a month does not correlate to the time of the year (no proof included), or the number of rain-days in the month (no proof included). As a consequence, the average evaporation on the dry days in the month is for most historical months higher than that for the rain-days, which is illustrated by the scatter diagram of Figure 10.3. There may be a slight positive trend between the difference $\overline{E_{pan,dry}} - \overline{E_{pan,r}}$ as a function of monthly rainfall, but the large scatter makes this trend insignificant. The average of all monthly differences $\overline{E_{pan,dry}} - \overline{E_{pan,r}}$ is 1.4 mm/day.

The variation within months between average pan evaporation on dry and on rain-days shows no spatial trend and little spatial variability in Zimbabwe. The average difference for all stations over the period 1995-1998 is 2.2 mm/day, which is greater than the 1.4 mm/day that was found as a 20-year average for Grasslands. The Grasslands average for 1995-1998 is also higher than the 20-year average, namely 1.87 mm/day. However, some unexplained inconsistency is found between the two databases, because the pan evaporation data for the period in which the databases overlap do not coincide on all days.

As mentioned, the possible trend in pan evaporation on dry and rain-days as a function of monthly rainfall is not significant. Moreover, for reasons of analytical tractability a constant difference is preferred. The difference is constant for Zimbabwe and with time. The real average difference is probably somewhere between 1.4 mm/day, as given by the long time series from Grasslands, and 2.2 mm/day, as shown by the database for all meteorological stations. For further theoretical derivation the absolute figure is not relevant. For further computations $\Delta E_{pan} = 2$ mm/day will be used for the whole of Zimbabwe.

These conclusions offer a practical improvement in the estimate of pan evaporation on dry and rain-days in months when only monthly data are available. The expected average pan evaporation on the rain-days within a month $\overline{E_{pan,r}}$ (mm/month) equals

$$ \mathrm{E}\left(\overline{E_{pan,r}} \,\middle|\, E_{pan,m}, P_m\right) = \frac{E_{pan,m} - \Delta E_{pan} * (n_m - n_r)}{n_m} \qquad \text{Eq. 10.1} $$

where

$E_{pan,m}$	is monthly measured pan evaporation	(mm/month)
ΔE_{pan}	constant, average difference between pan evaporation on dry and rain-days	(mm/day)
n_m	number of days in a month	(days)
n_r	number of rain-days in a month	(days)

If the number of rain-days in a month n_r is not measured, the Markov-derived estimate of Eq. 4.4 can be used.

$$ n_r = n_m \frac{p_{01}}{1 - p_{11} + p_{01}} \qquad \text{(days/month)} \qquad \text{Eq. 10.2} $$

The transition probabilities for a rain-day after a dry day p_{01} and for a rain-day after a rain-day p_{11} are a logistic function or a power function of the monthly rainfall P_m, see Eq. 3.7 and 3.8. Logically, the average pan evaporation on dry days is simply the average pan evaporation on rain-days plus the average difference;

$$ \mathrm{E}\left(\overline{E_{pan,dry}} \,\middle|\, E_{pan,m}, P_m\right) = \mathrm{E}\left(\overline{E_{pan,r}} \,\middle|\, E_{pan,m}, P_m\right) + \Delta(E_{pan}). $$

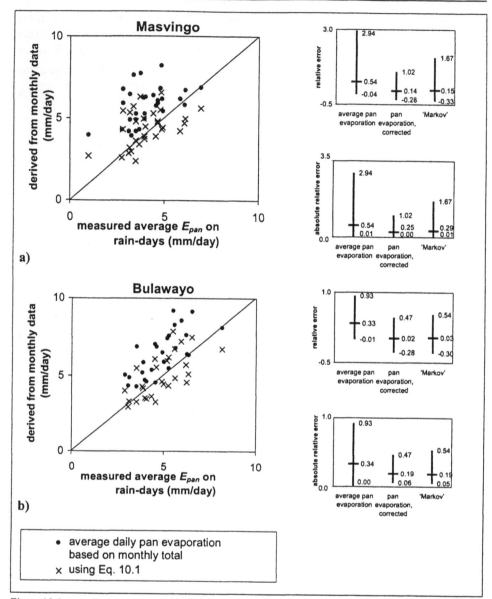

Figure 10.4   Comparison of different methods to derive average pan evaporation on rain-days from monthly data for a) Masvingo and b) Bulawayo. Here an average difference of 2 mm/day between pan evaporation on dry and rain-days is used. For each location, in the left diagram the average measured pan evaporation on rain-days is plotted against the daily average determined from monthly pan evaporation (•) and against the outcome of Eq. 10.1 (×). The diagrams on the right show respectively the relative error and the absolute relative error for the different methods. The maximum, average and minimum errors are marked. 'Pan evaporation, corrected' uses the historical number of rain-days, while 'Markov' uses the number of rain-days derived from the Markov chain.

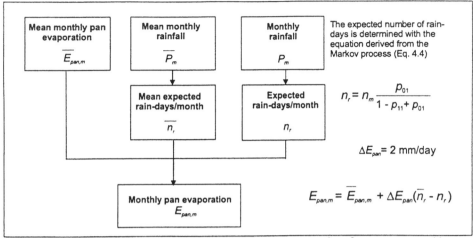

Figure 10.5   Determination of expected monthly pan evaporation from mean monthly pan evaporation, mean monthly rainfall and monthly rainfall.

Figure 10.4 shows that for Masvingo and Bulawayo the estimate of the average pan evaporation on the rain-days in the month is considerably improved. By using the measured number of rain-days and an average difference in pan evaporation between dry and rain-days of ΔE_{pan} = 2 mm/day, the bias in the estimate has diminished considerably and the variance in the relative error is smaller.

For Masvingo the average absolute relative error decreases from 0.54 to 0.25, for Bulawayo from 0.34 to 0.19. Use of the Markov properties to determine the number of rain-days gives almost no loss in the average absolute relative error (0.29 for Masvingo, 0.19 for Bulawayo), but more variance in the error. The average relative error is a measure for the bias. It decreased from 0.54 to 0.15 for Masvingo and from 0.33 to 0.02 for Bulawayo. The difference between dry and rain-days ΔE_{pan} = 2 mm/day has been used, which is representative for the whole of Zimbabwe. For 1995 - 1998 the average difference between dry and rain-days for Masvingo and Bulawayo is higher (2.4 mm/day for both). For this reason, a bias remains in the approximation.

The methodology to derive average pan evaporation on rain-days can also be used to estimate more accurate monthly pan evaporation when only mean monthly pan evaporation, mean monthly rainfall and monthly rainfall data are available. Figure 10.5 shows how this is done. Figure 10.6 shows that a considerably improved estimate is obtained.

Following the line of argument presented above, the knowledge that rain-days have 2 mm/day less pan evaporation than dry days can also serve daily models when daily pan evaporation data are available at fewer locations than rainfall data. Such a situation occurs when rainfall is determined through Cold Cloud Duration (Rugege, 2001). The meteorological stations Henderson and Mutoko, each some 130 km away from the Grasslands station, were used to verify the hypothesis that an improved estimate can be obtained. The Grasslands pan evaporation data applied to Mutoko yield an R^2 of 0.44 if only the days on which at both locations either rain or no rain has been recorded are considered (417 days).

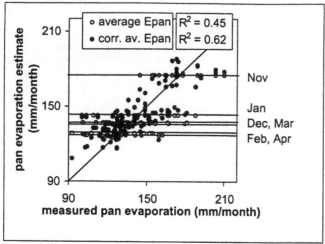

Figure 10.6   Scatter diagram of measured pan evaporation against average monthly pan evaporation (open dots) and against an estimate that uses average monthly pan evaporation and additionally monthly rainfall (closed dots). Data from Grasslands and Markov properties for Harare were used.

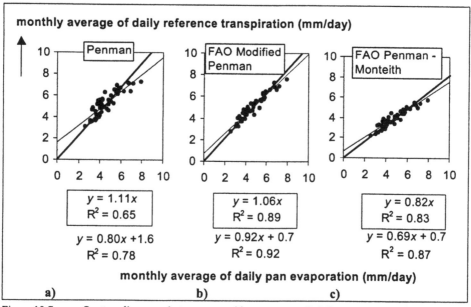

Figure 10.7   Scatter diagrams between monthly averages of daily pan evaporation and daily reference transpiration for three methods: a) Penman, b) FAO Modified Penman, c) FAO Penman-Monteith. Bold lines are linear regression models if intercept is set to zero. Slope and R^2 are given in box below. Thin line is linear regression model when intercept is not fixed. Slope and R^2 are given below, without a box.

For days on which one of the two stations records rain and the other does not, the R^2 is 0.03, thus correlation is negligible (146 days). If the Grasslands daily pan evaporation data are corrected by +2 mm/day if rain has occurred at Grasslands rain and not at Mutoko, and by -2 mm/day in the opposite situation, a slightly higher R^2 of 0.15 is obtained. For the combination Grasslands and Henderson, the values of R^2 are respectively 0.41 (190 days) and 0.14 (74 days). Thus, for Henderson there is no added benefit in the correction. The hypothesis is not really supported by the data. Unfortunately rainfall and pan evaporation data were not available for locations closer to each other. It is expected that in such a case the correction would have been more beneficial.

10.3 Reference potential evaporation

To translate pan evaporation into potential transpiration, a two-step method is normally used. First pan evaporation is related to a reference potential evaporation. Subsequently the reference potential evaporation is multiplied by a crop factor, to yield the potential dry-leaf transpiration of the crop.

The reference potential evaporation can be determined in various ways, yielding different estimates. Most widely used is the FAO Modified Penman method (Doorenbos & Pruitt, 1977), for which crop factors are extensively available. In 1990 (Smith, 1991) FAO adopted a new reference, the FAO Penman-Monteith potential evaporation, which refers to potential dry-leaf transpiration of a hypothetical crop similar to grass. For specifications, see Appendix C. In Figure 10.7 scatter diagrams of the Penman method, the FAO Modified Penman method and the FAO Penman-Monteith method are shown. In the following text only the FAO Penman-Monteith method is used, which is the latest reference. Figure 10.7 shows how for Zimbabwe, or at least for Grasslands, the transformation between different determinations of reference transpiration can be achieved. Makkink's method (1957, see textbooks e.g. De Laat, 1997) was not used here, as it only takes into account global radiation and temperature data. (For details on equations, see Appendix C.)

For the monthly average values of pan evaporation and the monthly average of daily reference transpiration, the 0-hypothesis of an intercept of 0 can be maintained. However, the 0-hypothesis does not hold true if regression is applied on the daily models (see Figure 10.8).

Figure 10.9 presents the relationship between pan evaporation and the FAO Penman-Monteith equation differently; the averages for the rain-days and dry days within one month are plotted. The scatter diagram shows that the relationship between pan evaporation and reference crop dry-leaf evapotranspiration is the same for averages of dry and rain-days. Also, the intercept can be set to zero and the slope to 0.82, regardless of occurrence of rain. Taking into account that the chicken wire on top of the pan reduces pan evaporation by some 10% (Torrance, 1981), the factor 0.82 is realistic. The estimation $E_{pot,ref,penmon} = 0.82 \, E_{pan}$ is used in further derivations.

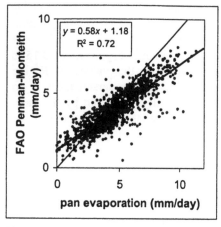

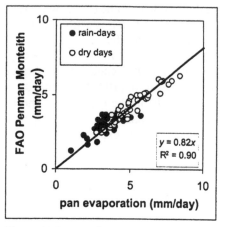

Figure 10.8   Scatter plot of daily pan evaporation and daily FAO Penman-Monteith evaporation.

Figure 10.9   Scatter plot of daily pan evaporation and daily FAO Penman-Monteith evaporation, both averaged over the dry and the rain-days in one month.

10.4   Penman-Monteith potential transpiration for different types of vegetation cover

For estimation of potential interception and transpiration from a vegetated surface, the Penman-Monteith equation is physically the best. Modifications have been proposed for theoretical and practical reasons, which resulted in tens of versions of the formula (De Bruin, 1987). In Table 10-1 representative vegetation parameters for grass, arable crops and trees are presented, as compiled from the literature by De Laat (1997) and Ward & Robinson (1990). Neither source gives an indication of the variance of the different parameters for a certain class. The crop parameters albedo r and aerodynamic resistance r_a, and, in the case of transpiration, crop resistance r_c, can usually not be determined with sufficient accuracy. Here the standard Penman-Monteith equation will be employed for the different vegetation types in order to obtain an indication of the size of the 'crop' factors k_c, which are used in two-step methods.

Crop factors k_c as referred to in the literature are ratios of the potential dry-leaf transpiration of the vegetation to the reference potential transpiration. In the Penman-Monteith equation potential transpiration relates linearly to the difference between potential interception and actual interception. Therefore, it is logical to first determine the potential interception, which in this case refers to the energy potential of evaporation from a wet canopy.

The Penman-Monteith equation for evaporation from a wet crop is similar to the Penman equation for an open water surface, with different values for the net radiation and the aerodynamic resistance.

$$I_{pot} = \frac{C}{L} \frac{sR_N + c_p \rho_a (e_a - e_d)/r_a}{s + \gamma}$$

Eq. 10.3

Table 10-1   Parameter values used for different types of vegetation cover. These values were compiled from the literature by De Laat (1997) and Ward & Robinson (1990). $U_{1.8}$ stands for the windspeed at a height of 1.8 m above the surface. Parameters in italics are not used for further computation.

symbol	parameter	unit	grass	arable crops	trees	open water
r	albedo	-	0.22	0.2	0.1	0.05
r_a	aerodynamic resistance, if no windspeed available	s/m	60	35	10	187
	if windspeed $U_{1.8}$ available		$\dfrac{\ln\left(\dfrac{1.8-\tfrac{2}{3}h_c}{0.123h_c}\right)\ln\left(\dfrac{1.8-\tfrac{2}{3}h_c}{0.0123h_c}\right)}{0.41^2 U_{1.8}}$		10	$\dfrac{245}{0.45U_2+0.5}$
h_c	height	m	0.15	1	n.a.	n.a.
r_c	min. crop resistance	s/m	70	30 (potatoes)	150	n.a.
			65	*80 (maize)*	*80*	
D_v	daily interception threshold of canopy	mm/ day	1	1	2	n.a.

For an explanation of the parameters, see Appendix C. The influence of the land cover is through the aerodynamic resistance r_a (s/m) and through the albedo r (-), which influences the net radiation R_N. Differences between dry and rain-days occur via actual temperature, sunshine hours and dewpoint pressure (humidity). As a consequence, differences in meteorological circumstances between dry and rain-days act differently upon potential interception for different land covers. The relations are non-linear (in the systems theory sense).

Figure 10.10a depicts the monthly averages of the daily Penman-Monteith potential interception values I_{pot}. They are plotted against monthly averages of daily pan evaporation data E_{pan}. Figure 10.10a shows that I_{pot} is proportional to E_{pan}. By regression proportionality coefficients were determined: 1.1 for grass, 1.9 for crops and 3.6 for trees.

Subsequently, the potential transpiration is determined

$$T_{pot} = \frac{s+\gamma}{s+\gamma(1+r_c/r_a)}(I_{pot}-I_{act}) \qquad \text{Eq. 10.4}$$

For an explanation of parameters, see Appendix C. The type of vegetation cover determines the crop resistance r_c and the aerodynamic resistance r_a in the ratio. The aerodynamic resistance r_a is also a parameter in the equation for I_{pot}, as is the albedo r, which influences the net radiation R_N. Differences between dry and rain-days occur via I_{pot}, I_{act} and s. For the determination of crop factors I_{act} is assumed 0, independent of whether rain occurred or not. The parameter s is the slope of the saturation vapour pressure-temperature curve, which is non-linearly dependent on the temperature.

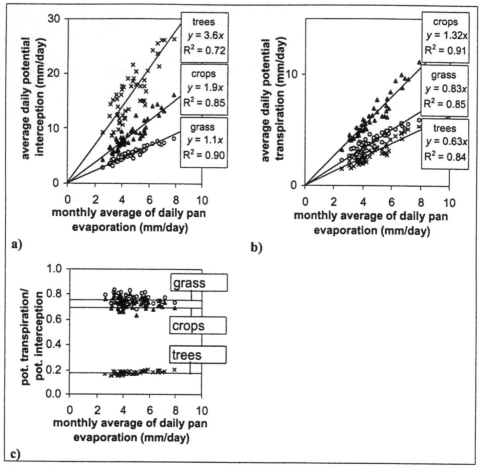

Figure 10.10   Scatter plots of monthly average of daily pan evaporation against monthly average of daily Penman-Monteith potential a) interception b) transpiration, using data from Grasslands (1995-1998). Linear models have been fitted, for which the intercept was set to 0. In c) monthly averages of the daily ratios of potential transpiration to potential interception are depicted.

For Grasslands the average s is 0.123. The average difference in s between dry and rain-days is 0.04 and the standard deviation is 0.02. As a result, the variable s and thus the factor $(s+\gamma)/(s+\gamma (1+r_c/r_a))$ are not significantly different on dry and rain-days. Therefore, transpiration is different on dry and rain-days through the different values for potential and actual interception.

The factor $(s+\gamma)/(s+\gamma (1+r_c/r_a))$ is in the order of 0.7 to 1 for grass and crops and around 0.2 for trees. Assuming canopy interception thresholds D of 1 mm/day for crops and grass, the influence of actual interception on potential transpiration is in the order of 0.7-1 mm/day. For trees interception thresholds are assumed 2 mm/day, which makes the influence 0.4 mm/day. The relative reduction in potential transpiration is 15 to 20 %.

As with Figure 10.10a for interception, Figure 10.10b shows the monthly average of the daily Penman-Monteith potential transpiration. For both dry and rain-days, I_{act} is set to 0, as the objective is to determine crop factors, which refer to dry-leaf transpiration. The proportionality coefficients that can be determined through regression are the crop factors k_c: 0.83 for grass, 1.32 for crops and 0.63 for trees. Thus, for trees the potential transpiration is less than for crops and grass![1]

As with pan evaporation, the differences between dry and rain-days were determined for the other recorded meteorological parameters. Figure 10.11 shows that for the monthly averages of the daily records of relative humidity, maximum temperature, minimum temperature, insolation and sunshine hours, similar conclusions to those for pan evaporation can be drawn:

- significant differences between dry and rain-days exist,
- there is a large scatter in the differences between the averages of dry and rain-days within a month,
- no significant trend can be determined between monthly rainfall and differences between averages of dry and rain-days.

Therefore, an average difference between dry and rain-days has been determined for all parameters, through minimisation of the residual sum of squares.

As shown in Figure 10.11, average differences between dry and rain-days are: for relative humidity -14% (rain-days have higher humidity), for maximum temperature during the day ΔT_{max} = 2.4 °C, for minimum temperature ΔT_{min} = - 0.9 °C, for insolation ΔR_S = 77 W/m^2, and for sunshine hours Δn = 4.3 hours/day. Unganai (1997) found significant negative correlation between surface temperatures and rainfall occurrence in Bulawayo and Harare.

With the approach that was used for pan evaporation (Eq. 10.1) and using average monthly meteorological parameters (Torrance, 1981) and the average number of rain-days per month (derived from rainfall database), the Penman-Monteith equation can be calculated. It is realised that this is not entirely correct, because using the mean parameters in a non-linear equation does not necessarily give the mean solution. However, because only just 3 years of daily data are available and the purpose is to study seasonal differences, substitution of the long-term mean values is the best option. The influence of considering different meteorological circumstances for dry and rain-days is presented in Figure 10.12, by comparing of a-d with e-h. Figures 10.13-10.15, show results of Figure 10.12 in a way that demonstrates the relative differences better. During the rainy season the absolute difference in potential transpiration between dry and rain-days is quite constant: 2.3 mm/day for grass, 3.1 for arable crops and 1.7 for trees. Trees have least potential transpiration. The relative difference between dry and rain-days has a higher variability than the absolute differences, but is in the range of 40 to 60% for all types of vegetation. The ratio of potential transpiration to potential interception also barely varies over the rainy season. Values are in the order of 74% for grass, 80% for arable crops and 18% for trees.

---

[1] In local studies crop factors still refer to FAO Modified Penman reference evaporation. However, FAO has adjusted its policy in 1990 and now refers to FAO Penman-Monteith reference evaporation, see Appendix C. In the programme CROPWAT, which can be downloaded from the internet (Clarke et al., 1998), crop factors k_c are presented for the different growing stages of many crops.

In Figure 10.10c a different ratio between potential transpiration and potential interception was presented for crops: 69% instead of 80%. For Figure 10.10c Grasslands windspeed data were used to compute aerodynamic resistance r_a for each day. For crops of 1 m in height the average of the resulting values of r_a is 27.5 s/m, rather than the 35 found in the literature which was used for Figure 10.12. This explains the difference. For grass the average Grasslands value of r_a is 80 s/m rather than 60 s/m, but due to non-linearity the ratios of potential transpiration to potential interception is for both 75%. In any case the parameters are only indicative. The height of the crop can easily vary, to say nothing about the aerodynamic and crop resistance, which are both difficult to determine.

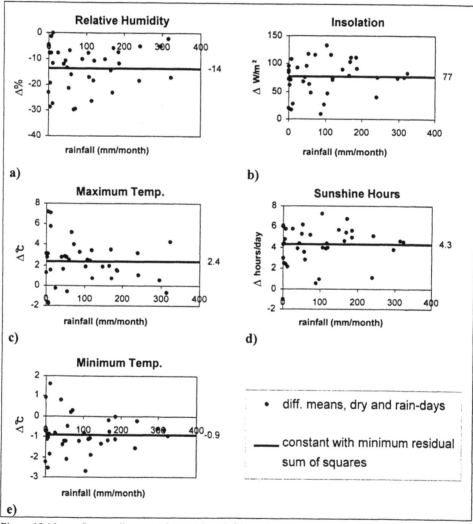

Figure 10.11   Scatter diagram of monthly rainfall against the difference between the averages of certain parameters for the dry days and the rain-days within the month. The parameters are a) relative humidity, b) insolation, c) maximum day temperature, d) number of sunshine hours in the day, c) minimum day temperature.

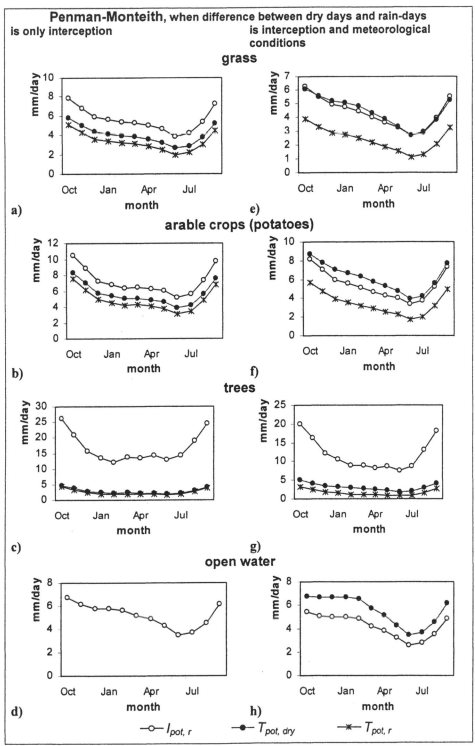

Figure 10.12   Penman-Monteith potential transpiration for different months in the year and for different types of land cover using average monthly parameters as input. Figures e-h show charts for the same classes as a-d, but now a distinction has been made for the meteorological parameters on dry and rain-days. The vegetation parameters used are shown in Table 10-1.

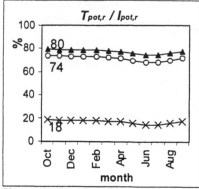

Figure 10.13   Potential transpiration on rain-days as a percentage of potential interception, for average meteorological circumstances in Harare, for different types of vegetation. Thus, actual interception is set to 0.

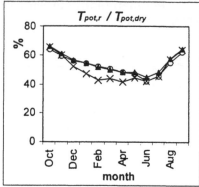

Figure 10.14   Potential transpiration on rain-days as a percentage of potential transpiration on dry days, for average meteorological circumstances in Harare, for different types of vegetation. Actual interception on rain-days is set to 1 mm/day for crops and grass and 2 mm/day for trees.

Figure 10.15   Difference between potential transpiration on dry days and on rain-days, for average meteorological circumstances in Harare, for different types of vegetation. Actual interception is set to 1 mm/day for crops and grass and 2 mm/day for trees.

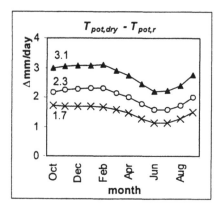

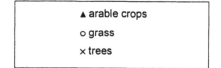

10.5   Average monthly potential transpiration

In the previous Sections it was shown that there is a considerable difference between potential transpiration on dry and on rain-days. The monthly transpiration model in this dissertation needs the monthly average of daily potential transpiration as input. It is beneficial to use the number of rain-days as estimated with the Markov process to determine the monthly potential transpiration from monthly pan evaporation, if interception is significant. This Section explains how monthly potential transpiration estimates improve. Figure 10.16 shows the procedure to be followed.

Thus, first the monthly pan evaporation is attributed to the expected number of rain-days and dry days, following Eq. 10.1. This yields an estimate of the average pan evaporation during the rain-days in the month and successively of the dry days in the month. For the dry days, the procedure is straightforward. The estimated average pan evaporation is multiplied by a pan factor to obtain the potential dry-leaf transpiration of a reference crop. Subsequently crop factors are applied and the potential dry-leaf transpiration of the crop is obtained.

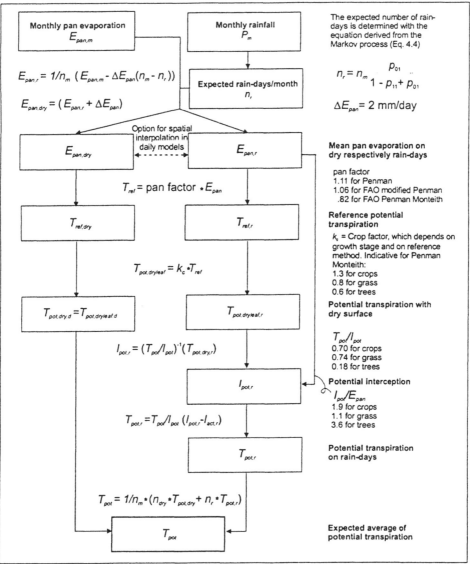

Figure 10.16   Flow chart of procedure to determine expected average potential transpiration. The given numbers are only indicative for Zimbabwe and were discussed in the previous Sections.

For rain-days, the estimated average pan evaporation is also multiplied by the same pan factor and the same crop factor. This yields the dry-leaf potential transpiration of the crop. However, the rain-day is not dry and therefore a correction needs to be applied for interception. To do this, first the potential interception is determined:

$$I_{pot,r} = \frac{s + \gamma(1 + r_c/r_a)}{s + \gamma} T_{pot,dryleaf,r} \qquad (-) \qquad \text{Eq. 10.5}$$

Subsequently the actual interception needs to be subtracted from the potential interception.

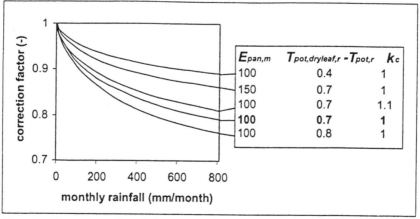

Figure 10.17   The ratio of the corrected estimate of the average daily potential transpiration to the average daily potential transpiration determined in the established way, plotted as a function of the monthly rainfall. The value of $T_{pot,r,dryleaf} - T_{pot,r}$ is varied for realistic values for crops and grass. The crop factor k_c and $E_{pan,m}$ influence the correction factor, but do not influence the absolute difference. To determine the number of rain-days, the Markov chain properties for Harare are used.

In Chapter 6 an estimate of the actual interception has been obtained. As interception is complementary to effective rainfall, following Eq. 6.24, the median of the average interception on rain-days is:

$$I_{act,r} = \beta - \beta \exp\left(\frac{-D_v}{\beta}\right)$$
(-)   Eq. 10.6

where, as defined before,

D_v is daily interception threshold of canopy   (mm/day)

β expected mean rainfall on rain-days in the month $= P_m/n_r$ (mm/day)

in which the expected number of rain-days n_r in the month is a function of the monthly rainfall P_m, see Eq. 4.4.

As a consequence, following Eq. 10.4, the expected potential transpiration during rain-days in the month is:

$$T_{pot,r} = \frac{s+\gamma}{s+\gamma(1+r_c/r_a)}\left(I_{pot,r} - I_{act,r}\right)$$
(-)   Eq. 10.7

Combining Eqs. 10.4-10.6 yields:

$$T_{pot,r} = T_{pot,r,dryleaf}$$
$$- \frac{s+\gamma}{s+\gamma(1+r_c/r_a)}\left(\frac{P_m}{n_r} - \frac{P_m}{n_r}\exp\left(\frac{-n_r D}{P_m}\right)\right)$$
(mm/day)   Eq. 10.8

When data are lacking the following estimates can be applied:

$$T_{pot,r,dryleaf} - T_{pot,r} = \frac{s+\gamma}{s+\gamma(1+r_c/r_a)}I_{act}$$
$$\approx 0.7 \text{ for crops and grass}$$
$$\approx 0.4 \text{ for trees}$$
(mm/day)   Eq. 10.9

This difference determines the impact that a distinction between dry and rain-days has on the estimate of average monthly potential transpiration. If $I_{act} = 0$, it is not necessary to follow the two paths in Figure 10.16, because they then both multiply and divide by the same factors. The established 'two-step method' to determine expected potential transpiration refers to dry-leaf potential transpiration only (e.g. applied in CROPWAT). The method presented in Figure 10.16 takes account of restricted transpiration on rain-days. Figure 10.17 presents the ratio of corrected average potential transpiration to average potential transpiration determined in the normal way. The absolute difference increases with monthly rainfall, with increases in $T_{pot,r,dryleaf} - T_{pot,r}$ and with increases in crop factor. The relative difference also depends on the monthly pan evaporation $E_{pan,m}$.

The difference $T_{pot,r,dryleaf} - T_{pot,r}$ is the maximum absolute difference between corrected average potential transpiration and the potential transpiration that is determined in the established way. This maximum difference occurs when the number of rain-days in the month n_r approaches the number of days in the month n_m. For Harare, according to the Markov process, this happens when monthly rainfall is 810 mm/month. The probability of exceedance of a monthly rainfall of 810 mm/month is negligible. Yet to show how the correction approaches the maximum, in Figure 10.17 the full range P_m from 0 to 810 mm/month is shown. For realistic values of rainfall (<500 mm/month), of crop parameters and of pan evaporation, correction factors range between 1 (no correction) and 0.9 or 0.8. A difference of 10 to 20% when monthly rainfall is high should not be ignored.

10.6 Closing remarks

The conclusion of the preceding Sections is that differences in potential transpiration between dry and rain-days are considerable, in the order of 40 to 60% for the wet season. A method is presented to assess, from monthly pan evaporation and monthly rainfall, separate averages of pan evaporation for the dry and the rain-days within a month. In this way an improved estimate of the monthly average of the daily potential transpiration is obtained (Figure 10.16). If only mean monthly pan evaporation data are available, the methodology also serves to determine better estimates of monthly pan evaporation (Fig. 10.5).

It should be noted that the use of pan evaporation for potential evaporation is severely criticised by Morton (e.g. 1995), who states that actual areal evaporation is related to pan evaporation in a complementary way; when actual evaporation is greater, pan evaporation is less. This is argued using field data for pan evaporation at different locations in relation to irrigated fields in desert areas. At the downwind end of the irrigated areas, pan evaporation is less, because the humidity is higher and the energy is consumed further upwind. Following Morton's view, the lower pan evaporation on rain-days can be the effect of higher actual evaporation on these days, making less energy available for pan evaporation, but without indicating lower potential transpiration. The academic discussion on this subject is ongoing.

The influence of meteorological circumstances on all forms of potential evaporation is predominant. However, hydrological models with daily time steps, but having only monthly evaporation data available, usually do not distinguish between dry and rain-days. There are two models that the author is aware of which make such a distinction:

- The SARR-model (US Army Engineer Division North Pacific, 1972) assumes that the ratio of potential transpiration on rain-days to that on dry days decreases linearly with rainfall amounts per day. The ratio reaches a minimum value of 10% for rainfall of 35 mm/day.

- A monthly rainfall runoff model developed by Haan (1972) and tested on small catchments in Kentucky uses Thornthwaite potential evaporation for potential evaporation on dry days. On rain-days the evaporation rate is reduced by a factor of 2, independent of the amount of rain.

Neither reference explains how this distinction between dry and rain-days was determined. As mentioned, in Zimbabwe there is no significant correlation between the amount of daily rainfall and the pan evaporation. The difference between pan evaporation on dry and rain-days is not proportional (as in Haan's model), but absolute. Apart from the advantages for monthly models, which are the topic of this dissertation, the methodology presented in the preceding Sections is potentially an improvement for hydrological models with daily time steps where only monthly evaporation data are available.

11 Spatial Interpolation of Daily Rainfall

11.1 Introduction

In Chapter 1 the advantages of using monthly instead of daily models were mentioned. Apart from advantages on the demand side (no interest in daily results, less processing work), it was noted that the lack of correlation between daily rain gauge data makes it difficult to obtain good estimates of areal precipitation. This Chapter quantifies the consequences of spatial averaging of daily rainfall for the estimation of monthly interception, for areas where there is little correlation between daily rainfall stations. It is proposed that a more realistic estimate of interception is obtained when the rainfall is averaged on a monthly basis and then used as input for the monthly interception model.

11.2 Rainfall occurrence derived by interpolation

In Section 1.4 it was shown that the correlation between two stations with daily rainfall series is often very low, due to the limited spatial scale of convective rainfall. Furthermore, it was shown that occurrence of point rainfall agrees with a Markov chain. To determine areal average daily rainfall for input into conceptual hydrological models, it is usual to obtain daily areal rainfall series through interpolation of daily records between stations in and around the area concerned. Spatial averaging methods include the Thiessen polygon, the inverse distance squared method and Kriging. Daily estimates of mean areal precipitation depend on the method used, but all methods are of the form:

$$\overline{P} = \sum_{i=1}^{n} a_i P_i \qquad \text{Eq. 11.1}$$

In this mean n gauges are considered, each with a weight a_i in the areal average. The various methods differ in their determination of the weight a_i.

We have seen that interception and transpiration depend on water availability in a non-linear way. This implies that the time average of the two types of evaporation is not identical for two separate rainfall series with the same average rainfall but with a different variability. Variability depends both on the occurrence of rain-days and on the amounts of rain on rain-days.

Imagine two rainfall stations for which rainfall occurrence has the Markov property and which have the same transition probabilities p_{01} and p_{11}. For analytical tractability the most extreme case is considered: there is no correlation in rainfall occurrence between the two stations. Regardless of the interpolation method used, whenever at least one of the two stations registers a rain-day, the interpolated series will also register a rain-day. In this way rainfall will occur more frequently in the interpolated time series than in the time series of either station. As a compensation factor, on average less rain falls on rain-days. The rainfall occurrence for the interpolated series again agrees to a Markov chain (illustrated by Figure 11.1).

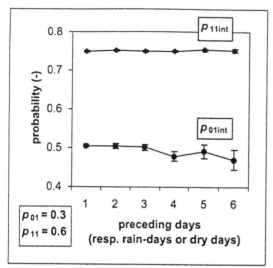

Figure 11.1   Probabilities of a rain-day after a
number of dry or rain-days, from an artificial time series
of 20,000 days that is derived by averaging two
stochastically-generated Markov chains for rain
occurrence. Both rain series use a $p_{01} = 0.3$ and $p_{11} =$
0.6.

Fiering (1997) mentions that if a stream is formed by the confluence of two streams in
which the stream flow agrees to a Markovian process, then in general the higher order
stream is not agreeing to a Markovian process. The lagged serial correlation of the
higher order stream is a function of the spatial relationship of the two lower order
streams. Thus, the Markov chain property of the interpolated series relies on the fact
that the two rainfall time series are not spatially correlated.

The question arises how the transition probabilities of the interpolated time series,
here named p_{11int} and p_{01int}, can be derived from the transition probabilities of the
separate stations' p_{01} and p_{11}.

Both stations have two states of a day, 0 (dry) or 1 (rain). For one station four
combinations of rain occurrence on two successive days are possible: 00, 01, 10, 11.
With 2 stations, 4*4 = 16 combinations are possible. The probability of occurrence for
each of these combinations can be determined. This is presented in Table 11-1 where
p is the probability that a rain-day will occur (Eq. 4.6). The sum of probabilities X_{00int}
is the probability that neither of the stations records rainfall on the two successive
days. The sum X_{01int} is the probability that neither of the stations records rainfall on
the first day and at least one records rainfall on the second day. And so on.

By substitution of $p = p_{01}/(1-p_{11}+p_{01})$, it can be found that $X_{01int} = X_{10int}$. As a result
the Markov transition probabilities for the interpolated case follow. For the
probability that the interpolated station has a rain-day after a dry day:

$$p_{01int} = \frac{X_{01int}}{X_{00int} + X_{01int}}$$ Eq. 11.2

Table 11-1   Probabilities of occurrence of rain on two successive days. The first two columns give the occurrence of rain on two successive days for two different stations. For instance, 01 means a dry day followed by a rain-day.

station 1	2	probability	the same letter indicates equal probabilities	parameter used for sum of probabilities
00	00	$(1-p)^2 (1-p_{01})^2$		X_{00int}
00	01		A	
01	00	$\}2(1-p)^2 p_{01}(1-p_{01})$		X_{01int}
01	01	$(1-p)^2 p_{01}^2$	B	
00	10		A	
10	00	$\}2p(1-p)(1-p_{11})(1-p_{01})$		X_{10int}
10	10	$p^2(1-p_{11})^2$	B	
11	11	$p^2 p_{11}^2$		
11	10		C	
10	11	$\}2p^2(1-p_{11})p_{11}$		
11	01		C	
01	11	$\}2p(1-p)p_{11}p_{01}$		X_{11int}
11	00			
00	11	$\}2p(1-p)p_{11}(1-p_{01})$		
10	01			
01	10	$\}2p(1-p)(1-p_{11})p_{01}$		

$$\frac{+}{1}$$

Numerator and denominator can both be divided by $(1-p)^2$. Consequently it appears that the denominator is equal to 1. In this way Eq. 11.2 transforms into:

$$p_{01int} = 2p_{01} - p_{01}^2 \qquad \text{Eq. 11.3}$$

Thus, p_{01int} is independent of p_{11}. Because $p_{01} < 1$, p_{01int} will always be greater than p_{01}, as expected. In Figure 11.2 the equation is depicted. Only for values of p_{01} close to 0 or close to 1, is p_{01int} close to p_{01}.

The probability of having a rain-day after a rain-day for the mean areal rainfall time series yields:

$$p_{11int} = \frac{X_{11int}}{X_{10int} + X_{11int}} \qquad \text{Eq. 11.4}$$

The result is depicted in Figure 11.3 for different values of p_{01} and p_{11}. The smaller p_{01} is, the more p_{11int} resembles p_{11}. But for a p_{01} of 0.1, the difference is already in the order of 0.03.

In analogy with Eq. 4.6, the probability that rainfall is recorded in the mean areal rainfall time series is:

$$P_{int} = \frac{p_{01int}}{1 - p_{11int} + p_{01int}} \qquad \text{Eq. 11.5}$$

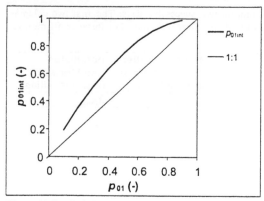

Figure 11.2 Relationship between p_{01} and p_{01int} independent of p_{11}.

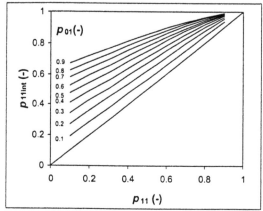

Figure 11.3 Relationship between p_{11} and p_{11int} for variable p_{01}.

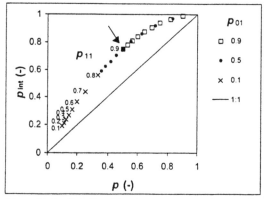

Figure 11.4 Relationship between p and p_{int} for variable p_{01} and p_{11}. For $p_{01} = 0.1$ (symbol x), the values of p_{11} are annotated. In a similar way as for $p_{01} = 0.1$, for $p_{01} = 0.5$ or 0.6 values of p_{int} increase with p_{11}. The value of p_{int} relates to p in the same way as p_{01int} does to p_{01}, see Figure 11.4. The arrow points to $p_{int} = 0.75$ which occurs for three combinations of transition probabilities ($p_{01} = 0.1$, $p_{11} = 0.9$), ($p_{01} = 0.5$, $p_{11} = 0.5$), ($p_{01} = 0.9$, $p_{11} = 0.1$).

By substitution of Eqs. 11.2 and 11.4 it can be shown that p_{int} relates to p in the same way that p_{01int} relates to p_{01}. In analogy with Eq. 11.3:

$$\boxed{p_{int} = 2p - p^2}$$

Eq. 11.6

The derivation is cumbersome, but the equation can easily be checked through substitution of values. Because $p < 1$, p_{int} will always be greater than p, as expected, with a maximum absolute difference of 0.25 and a maximum relative difference of 0.9. Thus, in comparison to the series from one station, the probability of rainfall occurrence can be almost twice as great for the mean areal rainfall series. Figure 11.4 shows that p increases with increasing values of p_{11} and p_{01}.

For p_{11int} a simple expression can also be derived, through the combining of Eqs. 11.4, 11.5 and 11.6:

$$p_{11int} = p_{01int}\left(1 - \frac{1}{p_{11int}}\right) + 1$$

$$= \left(2p_{01} - p_{01}^2\right)*\left(1 - \frac{1}{2p - p^2}\right) + 1$$

Eq. 11.7

11.3   Probability of exceedance

At a certain time step there are three possibilities: neither rain station records rain, one records rain, both record rain. The probabilities of these possibilities are simple:

Probability neither stations rain	one station rain	two stations rain
$(1-p)^2$	$2p(1-p)$	p^2

This means that for rain-days the probability that only one station has rain is:

$$P_{only1} = \frac{2p(1-p)}{2p(1-p) + p^2}$$

Eq. 11.8

Figure 11.5 shows how p_{only1} decreases with increasing values of p_{11} and p_{01}.

When only one station records rain, the interpolated time series records half of the recorded rain, because it is assumed that both stations have equal weight in the mean. Therefore, considering only rain-days for which rain is recorded at one station, the probability of exceedance is $1-F(P_{rint}|only\ 1) = exp(-2P_r/\beta)$. As a consequence, the probability of exceedance of rain on rain-days for the mean areal time series is:

$$1 - F\left(P_{rint}\right) = P_{only1}\,exp\left(\frac{-2P_r}{\beta}\right) + \left(1 - P_{only1}\right)exp\left(\frac{-P_r}{\beta}\right)$$

Eq. 11.9

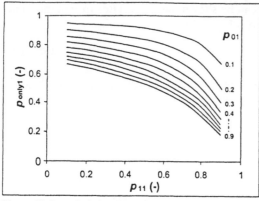

Figure 11.5   Relationship between p_{11} and the probability that only one of the two stations recorded rainfall p_{only1}. The relation is depicted for different values of p_{01}.

11.4   Consequences for interception and transpiration

Substitution of the above derived properties in the equations in Section 6.3 yields for the monthly interception of the mean areal series:

$$I_{m,\text{int}} = p_{\text{int}} * n_m \left\{ \begin{array}{l} - p_{only1} \dfrac{\beta}{2} \exp\left(\dfrac{-2D}{\beta}\right) \\[2mm] + \left(-\beta + p_{only1}\beta\right)\exp\left(\dfrac{-D}{\beta}\right) \\[2mm] + \beta - p_{only1}\dfrac{\beta}{2} \end{array} \right\}$$

Eq. 11.10

In Figure 11.6 it is illustrated that the use of the mean areal rainfall, determined from the equal weight of two rain stations with no correlation but with the same statistical characteristics, overestimates the interception by about 20% in the case of Zimbabwe.

Normally in daily hydrological models *a priori* estimates for the daily interception thresholds D are determined, depending on the vegetation cover. However, these estimates are fine-tuned by calibration. For monthly rainfall with $P_m < 300$ mm/month, Eq. 6.10 yields results that are similar to those of Eq. 11.10 if the interception threshold is lowered. For the example in Figure 11.6 this implies that when using the point interception equation, the interception threshold D is calibrated to 4 mm/day instead of the physical value 5 mm/day. This difference is less when correlation between rainfall stations exists, but overall it can be concluded that interception thresholds are underestimated if they are determined from the calibration of daily conceptual hydrological models that use mean areal rainfall estimates as input.

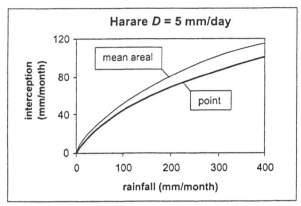

Figure 11.6   Relationship between rainfall and interception for point rainfall and for the mean areal rainfall, as specified in the text, with the use of the statistical characteristics of Harare and $D = 5$ mm/day.

If the threshold D is not calibrated, the interception may be overestimated by 20%. The example used only two rain stations. If more stations are used in the averaging process, the effect will be greater because the chance that all stations record a dry day decreases.

The overestimation of interception decreases effective rainfall, thus transpiration will be less. For months in the linear part of the monthly transpiration equation (Eq. 8.1), the absolute difference in transpiration is the product of slope B (see Eq. 8.10) and the absolute difference in interception. The total evaporation (interception + transpiration) will be overestimated by the product of the overestimation of interception ($0.2*I_m$) and ($1-B$). With a B of around 0.6, the overestimation of actual evaporation will amount to about $0.20 * 0.4 * I_m = 0.08 * I_m$ (mm/month). It depends on the value of the threshold D and on the value of S_{start} and P_m whether this amount is significant.

11.5   Correlation in occurrence of daily rainfall between rain stations

In the previous Sections it was determined how the use of mean areal rainfall would influence the estimate of monthly interception. The assumption was that there is no correlation between rainfall occurrence at the stations that are used in the analysis. Thus, an upper limit of the overestimation of interception was determined. Figure 1.2 showed that there is little correlation between daily rain stations. Near Harare the density of stations is comparatively high, but in more remote areas distances are easily in the order of 50 km, which would imply, according to Figure 1.2, a Pearson correlation in daily rainfall of 0.2 to 0.4.

For the analysis that was done in this Chapter, the correlation coefficient does not give much information. It is better to distinguish between correlation in rainfall occurrence and correlation in rain amounts on rain-days. In Table 11-2 transition probabilities of rainfall occurrence are presented. A distinction is made between the probability of rainfall occurrence that only depends on rainfall occurrence on the previous day at the same station (the figures in the diagonals) and probabilities that additionally depend on rainfall occurrence at a neighbouring station.

Table 11-2   Transition probabilities of Markov process, p_{11} and p_{01}, determined for the stations in the left column, under the condition that rainfall is recorded at the station in the upper row. For instance, the probability that rainfall will be recorded at Harare airport if it has rained at the same location on the previous day and if rain is recorded on the same day at Belvedere, is 0.86. The transition probabilities in the diagonals only depend on rainfall occurrence at one station. For Belvedere p_{11} is determined twice. The upper left p_{11} is for the period in which data from Harare Belvedere overlap with data from Harare airport resp. Marondera, the lower right p_{11} with Harare Kutsaga, resp. Chinhoyi. Airport, Belvedere and Kutsaga are all in Harare. Chinhoyi and Marondera are further away. See column 2 and Figure 11.7.

	distance to Belvedere (km)	this station records rain on second day ($\_1$)		
p_{11}		Airport	Belvedere	Kutsaga
Airport	13	0.60	0.86	
Belvedere	0	0.89	0.61	
			0.64	0.88
Kutsaga	16		0.84	0.53
p_{01}		Airport	Belvedere	Kutsaga
Airport		0.21	0.74	-
Belvedere		0.80	0.23	
			0.27	0.84
Kutsaga			0.80	0.33
p_{11}		Marondera	Harare Belvedere	Chinhoyi
Marondera	60	0.67	0.82	-
Harare Belvedere	0	0.80	0.64	
			0.64	0.79
Chinhoyi	75		0.80	0.55
p_{01}		Marondera	Harare Belvedere	Chinhoyi
Marondera		0.24	0.63	-
Harare Belvedere		0.67	0.27	
			0.26	0.62
Chinhoyi			0.71	0.35

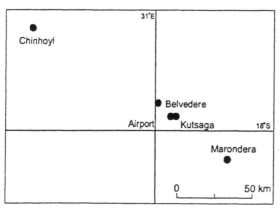

Figure 11.7 Map of rainfall stations used in Table 11-2.

Because the probabilities in the diagonals are considerably lower than the other probabilities, in particular in the case of p_{01}, it should be concluded that rainfall occurrence cannot be considered uncorrelated. This is still valid for long distances in the order of 75 km (Belvedere-Chinhoyi). For five-day rainfall anomalies, Makarau (1995) found Pearson correlation coefficients between Bulawayo and Lusaka (Zambia), West Transvaal (South Africa), Maun (Botswana) and the North West region of Madagascar to be as high as 0.6-0.7.

Apart from daily rainfall occurrence, daily rain amounts are also correlated. The correlation is more than can be expected from Figure 1.2. For the stations Harare Belvedere and Chinhoyi the daily rain is on average 10% more if rain is also recorded at the other station. The same is true for Harare Belvedere and Marondera.

For the above reasons, the errors made by using the mean areal rainfall as input for daily hydrological models are smaller than outlined in the previous Section. This shows that the Pearson correlation coefficient is a weak parameter to describe correlation, which is partly due to the influence of dry days.[23] Stol (1983) mentions that with convection as the main rainfall-producing mechanism, correlation between stations in zones on the scale of the subsidence zones can even become negative, while at longer distances the correlation is positive. The latter is due to large-scale convection-inducing mechanisms, in Zimbabwe the ITCZ (see Box 3-A).

The analysis in this Chapter is useful in the sense that it offers recommendations for stochastic daily rainfall generators that are based on the Markov chain. So far, stochastic rainfall models based on the Markov process have been mainly applied at points. To take spatial correlation into account, Bárdossy & Plate (1991, see also Wilby et al., 1994) applied semi-Markov processes on a number of atmospheric circulation patterns. Each circulation pattern has a different probability of occurrence of rainfall. Thus, for calibration, data on atmospheric circulation patterns needed to be processed. Pegram & Clothier (2000) suppose that a Markov process is valid at one point and that the extent and intensity of the rainfall event is determined through 'exponentiating a suitably scaled and shifted power-law filtered Gaussian random field'. They calibrated such models with the use of radar images. Table 11-2 suggests that spatial correlation can be introduced in a Markov stochastic rainfall model by making the probability of rainfall not only dependent on the occurrence of rainfall at the same station on the previous day, but also on the occurrence of rainfall on neighbouring stations on the previous day. (The probabilities in the table refer to occurrence of rainfall at other stations on the same day, but for numerical explicitness the previous day needs to be used.) The models by Bárdossy & Plate (1991) and Pegram & Clothier (2000) are more advanced and probably perform better. However, if only rainfall data are available, the solution that is suggested here may be promising.

---

[23] The Pearson correlation coefficient is the common correlation coefficient for rainfall on a daily basis. It assumes linear correlation between rainfall at stations and a normal distribution of the separate series. Particularly due to the occurrence of dry days, the correlation is not linear and daily rainfall is not normally distributed. Therefore the Pearson correlation coefficient may not be the most appropriate coefficient to correlate daily rainfall. Kendall's correlation coefficient, for example, is more robust for non-linear relationships (Hirsch et al., 1997).

The weakness of stochastic rainfall models lies often in the spatial distribution of rainfall. In this dissertation, monthly historical time series have been used. Such time series may not contain every possible rainfall distribution, but they at least contain spatial and time distributions that are feasible.

11.6 Closing remarks

The use of mean areal rainfall as input into a daily hydrological conceptual model means that interception is determined on the basis of rainfall series that have a higher probability of occurrence of rainfall and a lower average rainfall than point rainfall.

The arithmetic mean of two independent rainfall series with the same Markov properties yields a time series which also has a Markov property. For this time series, the transition probabilities p_{01int} and p_{11int} can be directly computed from the p_{01} and p_{11} of the individual stations. Additionally, distributions for the probability of exceedance of rain on rain-days can be derived, which compensate for the greater frequency of rain.

Subsequently, the effect on interception has been derived. Interception is either overestimated by some 20% or the threshold for daily interception is underestimated. The effect on overall evaporation is less significant, because with the increase in interception, effective rainfall and thus transpiration decreases. Although the Pearson correlation coefficients between series of daily rainfall from neighbouring stations are low, correlation exists in the occurrence of rainfall. Therefore the error made by using mean areal rainfall to determine interception is considerably less than 20%.

In many locations in the world the correlation between the different rainfall stations on a daily basis is little. By showing that systematic errors occur due to the use of mean areal rainfall as input in hydrological models on a daily basis, this Chapter presented quantitative evidence in favour of the use of monthly instead of daily models.

12 Conclusions

12.1 Synopsis of the methodology

The objective of this study was to improve monthly water balance models by establishing the relationship between monthly rainfall and the statistical characteristics of daily rainfall. This relationship is particularly important for the process of interception which has a time scale in the order of one day. Through interception, it also has a significant impact on the process of transpiration and surface runoff.

The statistical characteristics of daily rainfall (as a function of monthly rainfall) are not as variable in space as the daily or monthly rainfall itself. Determining such statistical characteristics at a few locations where sufficiently long daily records are available thus yields information on the daily variability of rainfall for larger areas. Monthly relationships between fluxes and stocks of water resources can be significantly improved by using this information.

Many authors have already studied the daily variability of rainfall. Their knowledge is incorporated in the stochastic models that generate daily rainfall records. However, because their aim was to develop a stochastic model, their attention was focused on describing daily rainfall variability as a function of the season rather than of the monthly rainfall. In this study the starting point was that if the monthly rainfall is high, then the chance that a rain-day is succeeded by a rain-day is also high. The probability that something will happen in the next time step conditional on what happens in the current time step, is the basis of stochastic models that use Markov processes. Compared to other stochastic methods, a Markov process has the great advantage that daily data are sufficient for calibration and validation and that all kinds of statistical characteristics of rainfall over longer time periods, in this case a month, can be derived relatively simply.

It has been shown in Chapter 3 that for Zimbabwe a Markov process of two states (dry/rainy) is sufficient to describe the variability within a month. This means that the Markov process depends on two transition probabilities only: the probability of a rain-day after a dry day, p_{01}, and the probability of a rain-day after a rain-day, p_{11}. The transition probabilities p_{01} and p_{11} can be described by logistic or power functions of the monthly rainfall. In particular p_{11} is spatially homogeneous. For many locations in the world where two state Markov processes have been used, it was shown that the following power functions between monthly rainfall P_m and the transition probabilities p_{01} and p_{11} are applicable.

$$p_{01} = q\left(P_m\right)^r \qquad\qquad (-) \qquad\qquad \text{Eq. 12.1}$$

$$p_{11} = u\left(P_m\right)^v \qquad\qquad (-) \qquad\qquad \text{Eq. 12.2}$$

where q, r, u and v are the only parameters that need to be calibrated at a few locations (~300 km) using daily rainfall series. Eqs. 12.1 and 12.2 correspond to Eqs. 3.9 and 3.11 respectively.

In theory power functions can surpass the value 1, however this does not happen for realistic values of monthly rainfall and calibrated values of q, r, u and v. Theoretically it is more correct to use a logistic function ($p_{01} = 1/(1+q\ P_m')$ and $p_{11} = 1/(1+u\ P_m')$, see Eq. 3.8), but this has not been done as it makes further derivations less transparent.

Making use of the mathematical properties of Markov processes and the research done by others, several equations have been derived that describe probability density functions of the variability of rainfall occurrence in a month.

Rainfall depths on rain-days can be described by an exponential probability density function, with a scale parameter β that is the mean rainfall on a rain-day (mm/day):

$$f(P_r) = \frac{1}{\beta}\exp\left(\frac{-P_r}{\beta}\right) \qquad\qquad \text{(-)} \qquad\qquad \text{Eq. 12.3}$$

The scale parameter β follows directly from the Markov process:

$$\beta = \frac{P_m}{n_r} = \frac{P_m(1 - p_{11} + p_{01})}{n_m p_{01}} \qquad\qquad \text{(mm/day)} \qquad\qquad \text{Eq. 12.4}$$

where n_r is the number of rain-days in a month and n_m is the number of days in a month (days/month). For climates where occurrence of rain-days does not agree to a Markov process, another relationship between the monthly rainfall P_m and the number of rain-days n_r can serve to compute β. Eq. 12.3 corresponds to Eq. 5.2 and Eq. 12.4 to Eq. 5.14.

Interception and transpiration

Interception on a daily basis is defined as the amount of daily rainfall that does not exceed a certain daily threshold. This daily threshold D depends on local land cover conditions. Using this definition, the monthly interception equation is:

$$I_m = P_m\left(1 - \exp\left(\frac{-D}{\beta}\right)\right) \qquad\qquad \text{Eq. 12.5}$$

Eq. 12.5 corresponds to Eq. 6.10 and yields the median of the monthly interception for a certain monthly rainfall. For any probability of exceedance, an analytical expression of the interception can be derived, see Eq. 6.13.

For transpiration, firstly the estimate of average potential transpiration has to be determined. For Zimbabwe it has been shown that pan evaporation differed on average 2 mm/day between dry and rain-days. Because the Markov process gives the probability density function of the number of rain-days, this difference can be used to improve estimates of monthly potential transpiration when only monthly pan evaporation and monthly rainfall data are available, see Figure 10.16. Depending on the land cover and the monthly rainfall, the ratio of the improved estimate to the established methods ranges between 0.8 and 1. If time series of monthly pan evaporation are not available but only seasonal averages, improved estimates of monthly pan evaporation can be determined, see Figure 10.5.

Transpiration is assumed to be potential T_{pot} as long as the available soil moisture content is more than a certain value S_b. When the soil moisture content is less than S_b, transpiration is constrained linearly.

Thus, at a certain point in continuous time the actual transpiration T_{act} is:

$$T_{act} = T_{pot} * \min\left(\frac{S}{S_b}, 1\right) \qquad \text{(mm/day)} \qquad \text{Eq. 12.6}$$

The maximum available soil moisture content is S_{max}. The ratio S_b/S_{max} is usually between 0.5 and 0.8, depending on land cover and soil type. Eq. 12.6 corresponds to Eq. 7.1, see also Figure 7.1.

The monthly transpiration equation which has been derived reads:

$$T_m(P_m) = \text{Min}\left(A + B * (P_m - I_m), T_{max,m}\right) \qquad \text{(mm/month)} \qquad \text{Eq. 12.7}$$

Eq. 12.7 corresponds to Eq. 8.1, see also Figure 8.1. The intercept A (mm/month) is the amount of transpiration obtained if there is no effective rainfall during the whole month. Depending on the value of the soil moisture at the start of the month $S_{start,m}$ two different equations for A apply:

$$A = S_{start,m}\left(1 - \exp\left(-\frac{1}{\gamma^\circ}\right)\right) \qquad \text{for } S_{start} < S_b \qquad \text{Eq. 12.8}$$

$$A = S_{start,m} - S_b \exp\left(-\frac{1}{\gamma^\circ} + \frac{S_{start,m}}{S_b} - 1\right) \qquad \text{for } S_{start} > S_b \qquad \text{Eq. 12.9}$$

where
γ° is a dimensionless parameter that represents the time it takes to transpire the soil moisture content S_b at the potential rate T_{pot}. Thus:

$$\gamma^\circ = \frac{S_b}{T_{pot,m} * \Delta t_m} \qquad \text{(-)} \qquad \text{Eq. 12.10}$$

where Δt_m is 1 (month) and $T_{pot,m}$ is the monthly potential transpiration. Eqs. 12.8 and 12.9 correspond to Eqs. 8.4 and 8.5. Eq. 12.10 corresponds to Eq. 8.2.

The slope B describes the change in monthly transpiration which results from a change in monthly effective rainfall. The slope obeys the following expression:

$$B = 1 - \gamma^\circ + \gamma^\circ \exp\left(-\frac{1}{\gamma^\circ}\right) \qquad \text{(-)} \qquad \text{Eq. 12.11}$$

The slope B follows from the assumption that effective rainfall is homogeneous during the month. It has been proven that this yields the expected monthly transpiration. Eq. 12.11 corresponds to Eq. 8.10.

The value of $T_{max,m}$ (mm/month) depends on the initial value of the soil moisture:

$$T_{max,m} = T_{pot,m} \qquad S_{start,m} > S_b \qquad \text{Eq. 12.12}$$

$$T_{max,m} = T_{pot,m} - \frac{S_b - S_{start,m}}{\Delta t_m}$$

$$- \left(P_{eff,m} - T_{pot,m} \right) * \gamma^\circ * \ln\left(\frac{S_b - \gamma^\circ * P_{eff,m} * \Delta t_m}{S_{start,m} - \gamma^\circ * P_{eff,m} * \Delta t_m} \right) \qquad S_{start,m} < S_b \qquad \text{Eq. 12.13}$$

where $P_{eff,m}$ is the effective monthly rainfall; P_m-I_m.

Additional equations

Monthly interception and transpiration are not the only monthly characteristics that are important for water resources management. Using the Markov properties and the exponential probability density function for rainfall depths on rain-days, several parameters have been derived that describe the variability of rainfall within the month. The number of rain-days n_r in a month with n_m days approaches a normal distribution with the following average and variance:

$$\text{E}(n_r|n_m) = n_m \frac{p_{01}}{1-C} \qquad (\text{-}) \qquad \text{Eq. 12.14}$$

$$\text{Var}(n_r|n_m) = n_m \frac{p_{01}}{1-C}\left(1 - \frac{p_{01}}{1-C}\right)\frac{1+C}{1-C} \qquad (\text{-}) \qquad \text{Eq. 12.15}$$

where $C = p_{11} - p_{01}$ and n_r and n_m are (in this case) counters of days, and thus dimensionless. Eqs. 12.14 and 12.15 correspond to Eqs. 4.4 and 4.5 respectively and have been used in the derivation of the monthly interception equation.

The probability density functions of the lengths of dry and wet spells (respectively n_{dry} and n_{wet}) are exponential equations:

$$\text{P}(n_{dry} = n) = p_{01} * (1 - p_{01})^{n-1} \qquad \text{Eq. 12.16}$$

$$\text{P}(n_{wet} = n) = (1 - p_{11}) * p_{11}^{n-1} \qquad \text{Eq. 12.17}$$

Eqs. 12.16 and 12.17 correspond to Eqs. 4.8 and 4.13. The probability density function of the length of dry spells is useful in the design of supplementary irrigation works. The probability density function of the length of wet spells can be used when planning rainwater harvesting methods.

The probability density function of the number of dry spells in a month is a Poisson distribution:

$$\text{P}\langle N_{dry}|n_m \rangle = \frac{\exp(-\lambda) * (\lambda)^{N_{dry}}}{N_{dry}!} \qquad \text{Eq. 12.18}$$

where N_{dry} is the number of dry spells and

$$\lambda = \frac{p_{01}}{1-C} * (1 - p_{11}) * n_m \qquad (\text{-}) \qquad \text{Eq. 12.19}$$

Eqs. 12.18 and 12.19 correspond to Eqs. 4.22 and 4.23. It is also shown that the probability density function of the number of wet spells is slightly different. The expected number of dry spells is equal to the expected number of pairs of a dry and wet spell. This information was used to derive the slope B in the monthly transpiration equation.

Often the length of the longest wet or dry spell influences the performance of rainfed agriculture more than the monthly rainfall. The longest dry spell determines crop destruction through water stress, the longest wet spell determines crop destruction because waterlogging occurs and the available energy is constraining transpiration. The cumulative density function which describes the probability that the length of the longest dry or wet spell ($n_{dry,max}$ resp. $n_{wet,max}$) in a month does not exceed a certain number of days n is an exponential distribution:

$$\mathbf{P}\left(n_{dry,max} \le n\right) = \exp\left[-n_m * \left(\frac{p_{01}}{1-C}\right) * (1-p_{11}) * (1-p_{01})^n\right] \qquad \text{Eq. 12.20}$$

$$\mathbf{P}\left(n_{wet,max} \le n\right) = \exp\left[-n_m * \frac{p_{01}}{1-C} * (1-p_{11}) * (p_{11})^n\right] \qquad \text{Eq. 12.21}$$

Eqs. 12.20 and 12.21 correspond to Eqs. 4.34 and 4.38 respectively.

In water resources models that compute crop yields to determine socio-economic performance, rainfed crops are planted at the start of the rainy season. If only monthly rainfall data are available, the expected date of the first rain-day within the month with the first rain is:

$$\mathrm{E}(n_{r,first}) = \left(p + (1-p)\frac{1}{p_{01}}\right)\Delta t_d \qquad \text{(days)} \qquad \text{Eq. 12.22}$$

where p is the probability of rain occurrence and $p = p_{01}/(1-C)$, see Eq. 4.6, and Δt_d is 1 (day). Eq.12.22 corresponds to Eq. 4.39.

Probability density functions for the total rain during a wet spell P_{wet} are useful to determine the risk of saturation, to determine the recovery of the soil moisture during wet spells, and to design rainwater harvesting systems. The cumulative density function is:

$$1 - \mathrm{F}(P_{wet}) = \exp\left(\frac{-P_{wet}(1-p_{11})}{\beta \Delta t_d}\right) \qquad \text{for } P_r > 0 \text{ (mm/day)} \quad (\text{-}) \qquad \text{Eq. 12.23}$$

Eq. 12.23 corresponds to Eq. 5.23.

Information requirements

Additional to the input monthly water resources models usually require, the methodology described in this dissertation makes use of daily rainfall data at a few locations in a region (~ 300 km × 300 km), to calibrate the model parameters q, u, r and v.

As with any monthly water resources model, input of monthly rainfall (mm/month) is necessary at a regional level (~ 50 km × 50 km). Monthly potential transpiration (mm/month) can be derived from pan evaporation data.

Furthermore, because the methodology in this dissertation makes a distinction between interception and transpiration, a daily threshold for transpiration D (mm/day) needs to be determined. Such a threshold depends on land cover, which is spatially heterogeneous. Land cover maps are necessary to determine D, but, as with daily models, a deterministic method to derive D from land cover maps is difficult.

Alternatively, the daily threshold D can be assessed through use of a monthly waterbalance model (see Section 6.2).

The maximum soil moisture content in the root zone S_{max} (mm) is a common parameter in monthly models. It depends on rooting depth and soil type. Like D, the parameter S_{max} is spatially heterogeneous and difficult to establish deterministically. The methodology assumes transpiration to be at a potential rate as long as the soil moisture is higher than S_b (mm). When data to determine S_b are lacking, S_b/S_{max} is assumed 0.5.

In Table 12-1 an overview of the information requirements is given.

Table 12-1   Model input.

parameters	description	information requirement
regional (~300 km)		
q, r, u, v	calibrated parameters which describe power relations between monthly rainfall P_m and transition probabilities p_{01} respectively p_{11}	daily rainfall data at a few locations in the region
subregional (~50 km)		
P_m (mm/month)	monthly rainfall	time series of rain gauge data
$T_{pot,m}$ (mm/month)	monthly potential transpiration time series	time series of pan evaporation; data on or at least seasonal averages of monthly pan evaporation
local (~1 km)		
D (mm/day)	daily interception thresholds	vegetation cover (and soil type); depending on leaf area and climate, typical values for D are $1 < D < 5$
S_{max} (mm)	maximum soil moisture content in root zone	vegetation cover and soil type
S_b (mm)	soil moisture content below which transpiration is water constrained.	vegetation cover and soil type $0.5 < S_b/S_{max} < 0.8$

12.2 Conclusions

General

- Markov transition probabilities (p_{01}, p_{11}) can be expressed as power functions or as logistic functions of monthly rainfall, which is a key to describe rainfall variability within the month for a region (~300 km × 300 km).

- With the use of the Markov process, probability density functions can be derived for the number of rain-days in a month, for the lengths of wet and dry spells, for the lengths of the longest wet and dry spells within the month, for the number of wet and dry spells in the month, and for the date of the first day with rainfall.

- For a certain monthly rainfall, the probability density function of rainfall depths on rain-days is an exponential function. The scale parameter of the exponential function β is the mean rainfall on a rain-day and can therefore be expressed as a function of the Markov transition probabilities and the monthly rainfall. For climates where rainfall occurrence is not according to the Markov process, another relationship between monthly rainfall P_m and the number of rain-days n_r is sufficient to express β as a function of monthly rainfall, because $\beta = P_m/n_r$.

- The properties of the exponential function imply that the scale parameter β is also the standard deviation in rainfall depth on rain-days and $\beta * \ln(2)$ is the median rainfall depth on rain-days.

- The high probability of occurrence of small amounts of rain is better represented by a mixed exponential distribution. However, this introduces two additional parameters that have to be calibrated. Only if all three parameters can be linked, for example by fixed ratios, can the mixed exponential distribution be expressed as a function of monthly rainfall.

- The βs derived by the Markov transition probabilities are performing at least as well in estimating average rain on rain-days as those derived by calibrating the exponential function to the cumulative probability density function of a class of monthly rainfall. The average rainfall on rain-days and the median rainfall on rain-days are underestimated by the calibrated β more severely than by the Markov-derived β. The variance in rain on rain-days is overestimated by the Markov-derived β to a greater extent than by the calibrated β.

- The determination of monthly interception from monthly rainfall data can be considerably improved. In the definition used here, interception is the amount of rainfall that does not exceed a certain daily threshold. For any probability of exceedance, interception can be expressed as a function of monthly rainfall. The median interception is the same as the mean interception and is the outcome of a simple exponential equation with β as a scale parameter.

- The equation for monthly interception is not particularly sensitive to spatial differences in the relationship between monthly rainfall and the transition probability p_{01}. Therefore derivation of statistical characteristics of rainfall occurrence at a few locations in the region only (spatial scale ~300 km) is sufficient.

- Existing monthly transpiration models do not agree with the generally accepted observation that at a certain point in time transpiration is potential as long as the soil moisture content is above a certain threshold, and is proportional to the soil moisture content when this is less than the threshold. Existing monthly transpiration models do not have effective rainfall contributing to the soil moisture during the month.

- The assumption of effective rainfall to be constant during the whole month, yields a simple analytical solution for the relationship between monthly effective rainfall and monthly transpiration. This relationship is linear for the lower range of monthly effective rainfall.

- A crucial parameter in the transpiration equation is the dimensionless time scale γ^p. It represents the time it takes to transpire the soil moisture content S_b at the potential rate T_{pot}.

- The monthly transpiration model yields transpiration estimates that are almost equal to the far more complex solution that disaggregates the month to wet and dry spells of expected lengths.

- Evaporation models can significantly be improved if a distinction between interception and transpiration is made. The equations for monthly interception and monthly transpiration are so simple that they can easily be applied in GIS models, without lengthening the computating time. The separation between interception and transpiration and the subsequent improvement of evaporation can improve the computation of monthly runoff. Moreover, it allows the separation of the water resources into blue water (surface water and groundwater), green water (transpiration) and direct feedback to the atmosphere (interception). Interception plays an important role in the persistence of rainfall locally, through its influence on the energy balance, and on a continental scale, through moisture recycling.

- Estimates of monthly potential transpiration from monthly rainfall and pan evaporation data can be improved by using the expected number of rain-days within the month, derived through the Markov process.

- The use of a daily threshold model for interception, applied on the average rainfall of two rainfall stations (which both agree to a Markov process and have similar statistical characteristics, but show virtually no correlation in rainfall occurrence) yields an overestimate of areal interception in the order of 20% as a result of the areal averaging. In cases with little correlation in daily rainfall occurrence, application of a monthly model for interception is better than a daily model which uses areal average rainfall as input.

- The methodology presented in this dissertation proves that with monthly data and a few time series of daily rainfall, monthly water resources models can be improved significantly. This offers a valuable tool to the water manager and the hydrologist, who need to make water resources planning decisions based on limited data and with limited available time. It is often more efficient for them to invest their time in large time series of monthly records, rather than in short time series of daily records. Improved monthly water resources models make it less urgent to invest in the collection and analysis of daily data, which requires denser networks.

- Instead of investing in stations that are gauged daily, it would be better to invest in a denser network that is gauged monthly, particularly if the occurrence of rain-days is also kept track of.

In relation to Zimbabwe

The following conclusions are valid for Zimbabwe and should be verified for other locations in the world.

- If the Markov process is described as a function of monthly rainfall, a first-order two state Markov process is adequate.
- The transition probability of a rain-day after a rain-day p_{11} is spatially more homogeneous than the transition probability of a rain-day after a dry day p_{01}. For all stations, $p_{11} = 0.20\ P_m^{0.24}$. For Harare, $p_{01} = 0.020\ P_m^{0.55}$, for Masvingo $p_{01} = 0.030\ P_m^{0.43}$ and for Bulawayo $p_{01} = 0.044\ P_m^{0.34}$.
- It is not necessary to make a distinction between months within the rainy season.

References

Abebe, B.B. (1996) Filtering the effect of orography from moisture recycling patterns. MSc thesis HH 268, IHE Delft.

Adane, A.A. (2000) Comparison of the hydrological models Aqua and Stream for the Zambezi river. MSc thesis DEW 136, IHE Delft.

Akaike, H. (1972) Information theory and an extension of the maximum likelihood principle. Second International Symposium on Information Theory, Akadémiai Kiadó. Budapest, Hungary.

Alley, W.M. (1984) On the treatment of evapotranspiration, soil moisture accounting and aquifer recharge in monthly water balance models. *Water Resources Research* 20: 1137-1149.

Almering, J.H.J., Bavinck, H., & Goldbach, R.W. (1988) Analyse, Delftse Uitgevers Maatschappij b.v., Delft, The Netherlands.

Bárdossy, A. (1998) Generating precipitation time series using simulated annealing. *Water Resources Research* 34: 1737-1744.

Bárdossy, A., & Plate, E.J. (1991) Modeling daily rainfall using a semi-Markov representation of circulation pattern occurrence. *Journal of Hydrology* 122: 33-47.

Bastiaanssen, W.G.M., Menenti, M., Feddes, R.A., & Holtslag, A.A.M. (1998) A remote sensing surface energy balance algorithm for land (SEBAL) 1. Formulation. *Journal of Hydrology* 212-213: 198-212.

Bastiaanssen, W.G.M., Pelgrum, H., Wang, J., Ma, Y., & Moreno, J.F. (1998) A remote sensing surface energy balance algorithm for land (SEBAL) 2. Validation. *Journal of Hydrology* 212-213: 213-229.

Bell, M., Faulkner, R., Hotchkiss, P., Roberts, N., & Windram, A. (1987) The use of dambos in rural development. Final Report to the Overseas Development Administration, UK Loughbourough University and University of Zimbabwe.

Benjamin, C.R., & Cornell, C.A. (1970) Probability, Statistics and Decision for Civil Engineers, McGraw-Hill, New York, USA. pp. 684.

Bierkens, M.F.P., Finke, P.A., & De Willigen, P. (2000) Upscaling and Downscaling Methods for Environmental Research, Kluwer Academic Publishers, Dordrecht, The Netherlands. pp. 190.

Binh, N.D., Murty, V.V.N., & Hoan, D.X. (1994) Evaluation of the possibility for rainfed agriculture using a soil moisture simulation model. *Agricultural Water Management* 26: 187-199.

Blöschl, G., & Sivapalan, M. (1995) Scale issues in hydrological modelling: a review. *Hydrological Processes* 9: 251-290.

Bo, Z., Islam, & S., Eltahir, E.A.B. (1994) Aggregation-disaggregation properties of a stochastic rainfall model. *Water Resources Research* 30: 3423-3435.

Buckle, C. (1996) Weather and Climate in Africa, Addison Wesley Longman Limited, Harlow, England. pp. 312.

Buishand, T.A. (1977) Stochastic modelling of daily rainfall sequences. Mededelingen Landbouwhogeschool Wageningen, Wageningen, The Netherlands.

Bullock, A. (1992a) The role of dambos in determining river flow regimes in Zimbabwe. *Journal of Hydrology* 134: 349-372.

Bullock, A. (1992b) Dambo Hydrology in southern Africa - review and reassessment. *Journal of Hydrology* 134: 373-396.

Calder, I.R., Harding, R.J., & Rosier, P.T.W. (1983) Objective Assessment of Soil-Moisture Deficit Models. *Journal of Hydrology* 60: 329-355.

Chin, E.H. (1977) Modelling daily precipitation occurrence process with Markov chain. *Water Resources Research* 13: 949-956.

Clarke, D., Smith, M., & El-Askari, K. (1998) CROPWAT for Windows: User Guide. FAO, Rome, Institute of Irrigation and Development Studies (IIDS), Southampton UK, National Water Research Centre (NWRC), Cairo, Egypt, http://www.fao.org/ag/AGL/aglw/cropwat.html.

Clarke, G.M., & Cooke, D. (1988) A basic course in statistics, Edward Arnold, London.

Clarke, R.T. (1998) Stochastic processes for water scientists: developments and applications, John Wiley and Sons, Chichester, England.

Cowperthwait, P.S.P. (1991) Further developments of the Neymann-Scott clustered point process for modelling rainfall. *Water Resources Research* 27: 1431-1438.

Cox, D.R., & Miller, H.D. (1965) The theory of stochastic processes, Chapman and Hall, London, United Kingdom.

De Bie, C.A.J.M. (2000) Comparative Performance Analysis of Agro-Ecosystems. PhD dissertation. Wageningen University and ITC Enschede, ITC publication No. 75, The Netherlands. pp. 232.

De Bruin, H.A.R. (1987) From Penman to Makkink. Evaporation and weather, Ede, The Netherlands. Proceedings and information No. 39. TNO Committee on Hydrological Research. The Hague, The Netherlands. pp. 5-16.

De Groen, M.M. (1999) Improving monthly time step water resources models; the case of interception. Sharing Scarce Resources: Land and Water Utilization in the Zambezi River Basin, Seminar II, Harare, Zimbabwe.

De Groen, M.M., & Savenije, H.H.G. (1996) Do land use induced changes of evaporation affect rainfall in Southeastern Africa? *Physics and Chemistry of the Earth* 20: 515-519.

De Groen, M.M., & Savenije, H.H.G. (1998) A Rainfall Generator for Southeastern Africa. *Physics and Chemistry of the Earth* 23: 399-403.

De Groen, M.M., & Savenije, H.H.G. (1999) Do land use-induced changes of evaporation affect rainfall? In *Water for Agriculture in Zimbabwe - Policy and Management Options for the Smallholder Sector*, Ed. Manzungu, E., Senzanje, A., Van der Zaag, P., University of Zimbabwe Publications, Harare, Zimbabwe. pp. 17-28.

De Laat, P.J.M. (1997) Soil-Water-Plant relations. Lecture Notes HH043/97/1, IHE Delft.

Department of Water Development / Water Resources Management Strategy (1998) Scoping the Updated Surface Water Resources Assessment of Zimbabwe (USWRAZ), summary of meeting held 15th January 1998 at DWD, Kurima House, Harare. internal report.

Dingman, S.L. (1998) Physical Hydrology, Prentice-Hall, New York. pp. 575.

Dooge, J.C.I. (1997) Scale problems in Hydrology, Chester C. Kisiel Memorial Lecture, University of Arizona. In *Reflections on Hydrology - Science and Practice* ed. Buras, Nathan, American Geophysical Union, Washington, U.S.A. pp. 85-143.

Doorenbos, J., & Pruitt, W.O. (1977) Crop water requirements. FAO Irrigation and Drainage Paper no. 24.

Eagleson, P.S. (1978) Climate, Soil and Vegetation - 4. The Expected Value of Annual Evapotranspiration. *Water Resources Research* 14: 731.

Eltahir, E.A.B. (1996) El Niño and the natural variability in the flow of the Nile River. *Water Resources Research* 32: 131.

Eltahir, E.A.B. (1998) A soil moisture-rainfall feedback mechanism 1. Theory and observations. *Water Resources Research* 34: 765-776.

Falkenmark, M. (1999) Forward to the future: A conceptual framework for water dependence. *Ambio* 28: 356-361.

Fiering, M.B. (1997) The real benefits from synthetic flows - reflections on 25 years with the Harvard Water Program, Chester C. Kisiel Memorial Lecture, University of Arizona. In *Reflections on Hydrology - Science and Practice* ed. Buras, Nathan, American Geophysical Union, Washington, U.S.A. pp. 17-33.

Filliben, J.J. (1975) The Probability Plot Correlation Test for Normality. *Technometrics* 17: 111-117.

Fitzpatrick, E.A., & Krishnan, A. (1967) A first-order Markov model for assessing rainfall discontinuity in central Australia. *Archeological Meteorology and Geophysical Bioklimatology* B15: 242-259.

Foufoula-Georgiou, E., & Georgiou, T.T. (1987) Interpolation of binary series based on discrete-time Markov chain models. *Water Resources Research* 23: 515-518.

Foufoula-Georgiou, E., & Lettenmaier, D.P. (1987) A Markov renewal model for rainfall occurrences. *Water Resources Research* 23: 875-884.

Gabriel, K.R., & Neumann, J. (1962) A Markov model for daily rainfall occurrence at Tel Aviv. *Quarterly Journal Royal Meteorological Society* 88: 90-95.

Gates, F., & Tong, H. (1976) On Markov chain modelling to some weather data. *Journal of Applied Meteorology* 15: 1145-1151.

Gbedzi, V.D. (1996) Selection of Subcatchments and Design of a Monitoring Network to Analyse the Influence of Land-use change on the Hydrology of the Mupfure Basin. MSc thesis HH 270, IHE Delft.

Gleick, P.H. (1987) The development and testing of a water balance for climate impact assessment: Modelling the Sacramento Basin. *Water Resources Research* 23: 1049-1061.

Gregory, J.M., Wigley T.M.L., & Jones, P.D. (1992) Determining and Interpreting the Order of a Two-State Markov Chain: Application to Models of Daily Precipitation. *Water Resources Research* 28: 1443-1446.

Gregory, J.M., Wigley, T.M.L., & Jones, P.D. (1993) Application of Markov models to area-average daily precipitation series and interannual variability in seasonal totals. *Climate Dynamics* 8: 299.

Grimmett, G.R., & Stirzaker, D.R. (2001) Probability and random processes, Third edition, Oxford Science Publications, Oxford, UK. pp. 596.

Haan, C.T. (1972) A water yield model for small watershelds. *Water Resources Research* 8: 58-69.

Haan, C.T., Allen, D.M., & Street, J.O. (1976) A Markov chain model for daily rainfall. *Water Resources Research* 12: 443-449.

Hall, M.J. (1996) Statistics and Stochastic Processes in Hydrology. Lecture Notes HH296/96/1, IHE Delft. pp. 194.

Hershenhorn, J., & Woolhiser, D.A. (1987) Disaggregation of daily rainfall. *Journal of Hydrology* 95: 299-322.

Hirsch, R.M., Helsel, D.R., Cohn T.A., & Gilroy, E.J. (1997) Statistical analysis of hydrologic data. In *Handbook of Hydrology* ed. Maidment, David R., McGraw-Hill, New York, USA.

Hoekstra, A.Y. (1998) Perspectives on water: an integrated model-based exploration of the future, International Books, Utrecht, The Netherlands.

Hoel, P.G., Port, S.C., & Stone, C.J. (1972) Introduction to Stochastic Processes: Houghton Mifflin Company, Boston, U.S.A.

Hughes, D.A. (1995) Monthly rainfall-runoff models applied to arid and semiarid catchments for water resource estimation purposes. *Hydrological Science Journal* 40: 751-770.

Hughes, D.A. (1997) Southern Africa "FRIEND" - The Application of Rainfall-Runoff Models in the SADC Region. WRC Report No 235/1/97, IWR Report No. 3/97, ISBN No. 1 86845 296 4, Water Research Commission South Africa.

Hutchinson, M.F. (1990) A point rainfall model based on a three-state continuous Markov occurrence process. *Journal of Hydrology* 114: 218-236.

Jimoh, O.D., & Webster, P. (1996) The optimum order of a Markov chain model for daily rainfall in Nigeria. *Journal of Hydrology* 185: 45-69.

Jones, P.G., & Thornton, P.K. (1997) Spatial and temporal variability of rainfall related to a third-order Markov model. *Agricultural and Forest Meteorology* 86: 127-138.

Katz, R.W. (1974) Computing probabilities associated with the Markov chain model for precipitation. *Journal of applied meteorology* 13: 953-954.

Katz, R.W. (1977) Precipitation as a chain dependent process. *Journal of Applied Meteorology* 13: 953-954.

Katz, R.W. (1981) On some criteria for estimating the order of a Markov chain. *Technometrics* 23: 243-249.

Kavvas, M.L., & Delleur, J.W. (1981) A stochastic cluster model of daily rainfall sequences. *Water Resources Research* 19: 1151-1160.

Kim, R. (1995) The water budget of heterogeneous soils - impact of soil and rainfall variability. PhD dissertation. Landbouwuniversiteit Wageningen. pp. 183.

Kreft, J. (1972) The distribution of thunderstorm days over Rhodesia. C.I.S.44, Rhodesia Meteorological Services, Harare, Zimbabwe.

Kupfuma, B., Nyengerai, A., Mupawose, R.M., & Ncube, H. (1992) Marketing and contribution of field crops to the national economy. In *Small-Scale Agriculture in Zimbabwe, Book one, Farming systems, policy and infrastructural development*, Ed. Whingwiri, E. E., Rukuni, M., Mashingaidze, K., Matanyire, C. M., Rockwood Publishers, Harare, Zimbabwe. pp. 86-116.

Kwadijk, J., & Van Deursen, W. (1999) Development and testing of a GIS based water balance model for the Rhine drainage basin. II-15, International Commission on the Hydrology of the Rhine (CHR/KHR), Lelystad, The Netherlands.

Laing, M.V. (1973) A detailed analysis of the normal rainfall of the Salisbury area. B47, Rhodesia Meteorological Services, Harare, Zimbabwe.

Lettau, H. (1969) Evapotranspiration Climatonomy I. A new approach to numerical prediction of monthly evapotranspiration, runoff and soil moisture storage. *Monthly Weather Review* 97: 691-699.

Lettau, H., Lettau, K., & Molion, L.C.B. (1979) Amazonia's Hydrological Cycle and the Role of Atmospheric Recycling in Assessing Deforestation Effects. *Monthly Weather Review* 107: 227-235.

Lettau, H.H., & Hopkins, E.J. (1991) Evapoclimatonomy III: The Reconciliation of Monthly Runoff and Evaporation in the Climatic Balance of Evaporable Water and Land Areas. *Journal of Applied Meteorology* 30: 776-792.

Lindström, G., Johansson, B., Persson, M., Gardelin, M., & Bergström, S. (1997) Development and test of the distributed HBV-96 hydrological model. *Journal of Hydrology* 201: 272-288.

Lørup, J.K., Refsgaard, J.C., & Mazvimavi D. (1998) Assessing the effect of land use change on catchment runoff by combined use of statistical tests and hydrological modelling: Case studies from Zimbabwe. *Journal of Hydrology* 205: 147-163.

Lu, Z.-Q., & Berliner, L.M. (1999) Markov switching time series models with application to daily runoff series. *Water Resources Research* 35: 523-534.

Lupankwa, M. (1996) The use of remote sensing in the study of dambos in Mashonaland East, Zimbabwe. Ext. Abst. of Conf. on Application of RS and GIS in Environmental and Natural Resources Assessment in Africa, Harare, Zimbabwe. Environment and Remote Sensing Institute, Harare, Zimbabwe. pp. 122-125.

Madamombe, E.K. (1994) A comparison of Neymann-Scott rectangular pulses and Markov chain stochastic rainfall models using data from Tanzania. MSc thesis, Department of Civil Engineering, University of Newcastle upon Tyne, United Kingdom.

Makarau, A. (1995) Intra-seasonal Oscillatory Modes of the Southern Africa summer circulation. PhD dissertation. University of Cape Town, South Africa. pp. 125.

Makhlouf, Z., & Michel, C. (1994) A two-parameter monthly water balance model for French watersheds. *Journal of Hydrology* 162: 299-318.

Makkink, G.F. (1957) Testing the Penman formula by means of lysimeters. *Journal Int. of Water Engineering* part A: 15-21.

Makurira, H. (1997) Integrated water management: Harare water supply study for Kunzwi and Manyame dams. MSc thesis DEW 011, IHE Delft.

Mare, A. (1998) 'Green water' and 'Blue water' in Zimbabwe: The Mupfure river basin case. MSc thesis DEW 044, IHE Delft.

Matarira, C.H. (1990) Drought over Zimbabwe in a regional and global context. *International Journal of Climatology* 10: 609.

Matarira, C.H., & Flocas, A.A. (1989) Spatial and temporal rainfall variability over Southeastern Central Africa during extremely dry and wet years. *Journal of Meteorology* 14: 3-9.

Matarira, C.H., & Jury, M.R. (1992) Contrasting Meteorological structure of intra-seasonal wet and dry spells in Zimbabwe. *International Journal of Climatology* 12: 165-176.

Mather, J.R. (1981) Using computed stream flow in watershed analysis. *Water Resources Bulletin* 17: 747-482.

Matola, J.R. (1998) Water resources planning approach for the Umbeluzi river basin. MSc thesis DEW 043, IHE Delft.

McCartney, M.P. (1998) The hydrology of a headwater catchment containing a dambo. PhD dissertation. University of Reading, United Kingdom. pp. 266.

Merka, J. (2000) Remote Sensing and GIS Applications to Water Resources Assessment and Management - A case-study in the Upper Mupfure River Catchment. MSc thesis, ITC, Enschede, The Netherlands.

Mimikou, M. (1984) Study for Improving Precipitation Occurrences Modelling with a Markov Chain. *Journal of Hydrology* 70: 25-33.

Ministery of Water Resources and Development (1984) An assessment of the surface water resources of Zimbabwe and guidelines for development planning. Harare, Zimbabwe.

Mood, A.M., Graybill, F.A., & Boes, D.C. (1963) Introduction to the theory of Statistics, McGrawHill, New York, USA.

Morton, F.I. (1983) Operational estimates of areal evapotranspiration and their significance to the science and practice of hydrology. *Journal of Hydrology* 66: 1-76.

Morton, F.I. (1995) Evaporation and feedback mechanisms in hydrology. In *Time and the River, Essays by Eminent Hydrologists* ed. Kit, Geoff ed., Water Resources Publications pp. 155-200.

Mudege, N.R. (1999) Supplementary irrigation for communal and small holder farming in Zimbabwe - Mupfure River Basin case - a preliminary viability study. MSc thesis DEW 094, IHE Delft.

Muir, K. (1994) Agriculture in Zimbabwe. In *Zimbabwe's agricultural revolution*, Ed. Rukuni, M., Eicher, C. K., University of Zimbabwe Publications, Harare, Zimbabwe. pp. 40-55.

Mulligan, M., & Reaney, S. (1999) Modelling storms from historic and GCM data for climate impact studies. poster presented at 24th General Assembly of the European Geophysical Society, The Hague, The Netherlands.

Musariri, M. (1998) Preliminary analysis for the Mupfure experimental catchments. MSc thesis HH 343, IHE Delft.

Nhidza, E. (1999) Integrating forestry land use in water resources management: Odzani river catchment case. MSc thesis DEW 089, IHE Delft.

Nicholson, S.E., Kim, J., Ba, M.B., & Lare, A.R. (1997) The mean surface water balance over Africa and its interannual variability. *Journal of Climate* 10: 2981-3002.

Nicholson, S.E., Lare, A.R., Marengo, J.A., & Santos, P. (1997) A revised version of Lettau's evapoclimatonomy model. *Journal of Applied Meteorology* 35: 1673-1676.

Nyagwambo, N.L. (1998) 'Virtual Water' as a demand management tool: the Mupfure river basin case. MSc thesis DEW 045, IHE Delft.

Nyamudeza, P. (1999) Agronomic practices for low rainfall regions in Zimbabwe. In *Water for Agriculture - Policy and Management Options for the Smallholder Sector* , Ed. Manzungu, E., Senzanje, A., Van der Zaag, P., University of Zimbabwe Publications, Harare, Zimbabwe. pp. 49-63.

Palmer, W.C. (1965) Meteorologic drought. Research Paper U.S. Weather Bureau, pp. 45.

Pegram, G.S., & Clothier, A.N. (2000) High resolution space-time modelling of rainfall: the 'String of Beads'-model. *accepted for Journal of Hydrology*

Penman, H.L. (1948) Natural evaporation from open water, bare soil and grass. Proceedings, Royal Society, 193. London. pp. 120-146.

Penman, H.L. (1949) The dependence of transpiration on weather and soil conditions. *Journal of Soil Sciences* 1: 74-89.

Pitman, W.V. (1973) A mathematical model for generating monthly river flows from meteorological data in Southern Africa. 2/73, University of Witwatersrand, Department of Civil Engineering, Hydrological Research Unit, South Africa.

Refsgaard, J.C., & Knudsen, J. (1996) Operational validation and intercomparison of different types of hydrological models. *Water Resources Research* 32: 2189-2202.

Rodriguez-Iturbe, I., Febres de Power, B., & Valdes, J.B. (1987) Rectangular pulses point process models for rainfall: analysis of empirical data. *Journal of Geophysical Research* 92: 9645-9656.

Roldán, J., & Woolhiser, D.A. (1982) Stochastic Daily Precipitation Models. 1. A Comparison of Occurence Processes. *Water Resources Research* 18: 1451-1459.

Rugege, D. (2001) Regional Analysis of Maize-based Land Use Systems for Early Warning Applications. concept PhD dissertation. Wageningen University and ITC Enschede, The Netherlands. to be published 2002.

Rukuni, M., & Makadho, J. (1994) Irrigation development. In *Zimbabwe's agricultural revolution*, University of Zimbabwe Publications, Harare, Zimbabwe. pp. 127-138.

Rushton, K.R., & Ward, C. (1979) The estimation of recharge. *Journal of Hydrology* 66: 345-361.

Rutter, A.J., Kershaw, K.A., Robins, P.C., & Morton, A.J. (1971) A predictive model of rainfall interception in forests. I. Derivation of the model from observations in a plantation of Corsican pine. *Agricultural Meteorology* 9: 267-384.

Savenije, H.H.G. (1995) New definitions for moisture recycling and the relationship with land use changes in the Sahel. *Journal of Hydrology* 167: 57-78.

Savenije, H.H.G. (1996) The runoff coefficient as the key to moisture recycling. *Journal of Hydrology* 176: 219-225.

Savenije, H.H.G. (1997) Determination of evaporation from a catchment water balance at a monthly timescale. *Hydrology and Earth System Sciences* 1: 93-100.

Savenije, H.H.G. (2000) Water Scarcity Indicators; the Deception of Numbers. *Physics and Chemistry of the Earth* 25: 199-204.

Savenije, H.H.G., & Van der Zaag, P. (2000) Conceptual framework for the management of shared river basins; with special reference to the SADC and EU. *Water Policy* 2: 9-45.

Schellekens, J., Scatena, F.N., Bruijnzeel, L.A., & Wickel, A.J. (1999) Modelling rainfall interception by a lowland tropical rain forest in northeastern Puerto Rico. *Journal of Hydrology* 225: 168-184.

Schreiber (1904) Über die Beziehungen zwischen dem Niederschlag und der Wasserführung der Flusse Mitteleuropa. *Meteorologische Zeitschrift.* 21 (10): 441-452.

Schwarz, G. (1978) Estimating the dimensions of a model. *Ann. Stat.* 6: 461-464.

Scoones, I., Chibudu, C., Jeranyama, P., Machaka, D., Machanja, W., Mavedzenge, B., Mombeshora, B., Mudhara, M., Mudziwo, C., Murimbarimba, F., & Zirereza, B. (1996) Hazards and opportunities - Farming livelihoods in dryland Africa: lessons from Zimbabwe, Zed Books Ltd, London, United Kingdom. pp. 267.

Şen, Z. (1976) Wet and dry periods of annual flow series. *Journal of the Hydraulic Division, ASCE* Proc. Pap. 12457, 102 (HY10): 1503-1514.

Şen, Z. (1977) Run-sums of annual flow series. *Journal of Hydrology* 35: 311-324.

Şen, Z. (1978) Autorun analysis of hydrologic time series. *Journal of Hydrology* 36: 75-85.

Şen, Z. (1980) Statistical analysis of hydrological critical droughts. *ASCE J. Hydraulic Division* 106: 99-115.

Seyam, I.M. (1999) Algorithms for water resources distribution in international river basins. MSc thesis DEW 073, IHE Delft.

Sharma, T.C. (1996a) A Markov-Weibull rain-sum model for designing rain water catchment systems. *Water Resources Management* 10: 147-162.

Sharma, T.C. (1996b) Simulation of the Kenyan longest dry and wet spells and the largest rain-sums using a Markov model. *Journal of Hydrology* 178: 55-67.

Shaw, E. M. (1988) Hydrology in Practice, Second Edition, Van Nostrand Reinhold (International), London, United Kingdom.

Shoniwa, S. (1996) Review of water resources, water use and water rights in the Mupfure basin. MSc thesis HH 277, IHE Delft.

Shuttleworth, W.J. (1997) Evaporation. In *Handbook of Hydrology* ed. Maidment, David R., McGraw-Hill, New York, USA. pp. 4.1-4.53.

Small, M.J., & Morgan, D.J. (1986) Relationship between a Continuous-Time Renewal Model and a Discrete Markov Chain Model of Precipitation Occurrence. *Water Resources Research* 22: 1422-1430.

Smith, J.A. (1987) Statistical Modeling of Daily Rainfall Occurrences. *Water Resources Research* 23: 885-893.

Smith, M. (1991) Report on the expert consultation on procedures for revision of FAO guidelines for prediction of crop water requirements. FAO, Rome, Italy. pp. 54.

Smit, G.N., & Rethman, N.F.G. (2000) The influence of tree thinning on the soil water in a semi-arid savanna of southern Africa, *Journal of Arid Environments* 44: 41-59.

Smith, R.E., & Schreiber, H.A. (1974) Point processes of seasonal thunderstorm rainfall 2. Rainfall depth probabilities. *Water Resources Research* 10: 418-423.

Stedinger, J.R., Vogel, R.M., & Foufoula-Georgiou, E. (1997) Frequency Analysis of Extreme Events. In *Handbook of Hydrology* ed. Maidment, David R., McGraw-Hill, New York, USA.

Steenhuis, T.S., & Van der Molen, W.H. (1986) Thornthwaite-Mather Procedure as a Simple Engineering Method to Predict Recharge. *Journal of Hydrology* 84: 221-229.

Stern, R.D., & Coe, R. (1982) The use of rainfall model in agricultural planning. *Agricultural Meteorology* 26: 35-50.

Stern, R.D., & Coe, R. (1984) A model fitting analysis of daily rainfall data. *Journal Royal Statistical Society A.* 147: 1-34.

Stewart, J. (1991) Econometrics, Philip Allen.

Stol, Ph.Th. (1983) Rainfall interstation correlation functions. VII. On Non-monotousness. *Journal of Hydrology* 64: 69-92.

Taylor, C.M., Said, F., & Lebel, T. (1997) Interactions between the land surface and mesoscale rainfall variability during HAPEX-Sahel. *Monthly Weather Review* 125: 2211-2227.

Thomas, H.A. (1981) Improved methods for National Water Assessment. contract WR15249270, U.S. Water Resources Council, Washington D.C., USA.

Thomas, H.A., Martin, C.M., Brown, M.J., & Fiering, M.B. (1983) Methodology for water resource assessment. report to U.S. Geological Survey, NTIS 84-124163, National Technical Information Service.

Thompson, S.A. (1992) Simulation of climate change impacts on water balances in the Central United States. *Physical Geography* 13: 31-52.

Thornthwaite, C.W., & Mather, J.R. (1955) The water balance. *Publications in Climatology, Drexel Institute of Technology, Laboratory of Climatoloty, New Jersey* 8: 1-104.

Thornthwaite, C.W., & Mather, J.R. (1957) Instructions and tables for computing potential evapotranspiration and the water balance. *Publications in Climatology, Drexel Institute of Technology, Laboratory of Climatoloty, New Jersey* 10: 185-195.

Thornton, P.K., Bowen, W.T., Ravelo, A.C., Wilkens, P.W., Farmer, G., Brock, J., & Brink, J.E. (1997) Estimating millet production for famine early warning: an application of crop simulation modelling using satellite and ground-based data in Burkina Faso. *Agricultural and Forest Meteorology* 83: 95-112.

Todorovic, P., & Woolhiser, D.A. (1975) A stochastic model for n-day precipitation. *Journal of Applied Meteorology* 14: 17-24.

Torrance, J.D. (1975) Availability of Atmospheric Water. *Transactions of the Rhodesian Scientific Association* 56: pp. 31.

Torrance, J.D. (1981) Climate Handbook of Zimbabwe. 551.582(689.1), Zimbabwe, Department of Meteorological Services, pp. 221.

Troch, P.A., Mancini, M., Paniconi, C., & Wood, E.F. (1993) Evaluation of a distributed catchment scale water balance Model. *Water Resources Research* 29: 1805-1817.

Tyson, P.D. (1986) Climatic Change and Variability in Southern Africa, Oxford University Press, Cape Town, South Africa. pp. 220.

U.S. Army Engineer Division, N.P.D. (1972) Program description, & user manual for SSAR-model, Streamflow synthesis and reservoir regulation. Program 724-k5-G0010, Portland, Oregon.

Unganai, L.S. (1996) Historic and future climatic change in Zimbabwe. *Climate Research* 6: 137.

Unganai, L.S. (1997) Surface temperature variation over Zimbabwe between 1897 and 1993. *Theor. Appl. Climatology* 56: 89.

Van de Giesen, N.C., Stomph, T.J., & De Ridder, N. (2000) Scale effects of Hortonian overland flow and rainfall-runoff dynamics in a West African catena landscape. *Hydrological Processes* 14: 165-175.

Vandewiele, G.L., Xu, C.-Y., & Ni-Lar-Win (1992) Methodology and comparative study of monthly water balance models in Belgium, China and Burma. *Journal of Hydrology* 134: 315-347.

Velghe, T., Troch, P.A., De Troch, F.P., & Van de Velde, J. (1994) Evaluation of cluster-based rectangular pulses point process models for rainfall. *Water Resources Research* 30: 2847-2858.

Vörösmatry, C.J., & Moore, B. (1991) Modeling basin-scale hydrology in support of physical climate and global biogeochemical studies: an example using the Zambezi river. *Surveys in Geophysics* 12: 271-311.

Ward, R. C., & Robinson, M. (1990) Principles of Hydrology, Third edition, McGraw-Hill, London.

Waymire, E.C., & Gupta, V.K. (1981a) The mathematical structure of rainfall representations, 1. A review of stochastic rainfall models. *Water Resources Research* 17: 1261-1272.

Waymire, E.C., & Gupta, V.K. (1981b) The mathematical structure of rainfall representations, 2. A review of the theory of point processes. *Water Resources Research* 17: 1273-1285.

Waymire, E.C., & Gupta, V.K. (1981c) The mathematical structure of rainfall representations, 3. Some applications of the point process theory to rainfall processes. *Water Resources Research* 17: 1287-1294.

Wilby, R., Greenfield, B., & Glenny, C. (1994) A coupled synoptic-hydrological model for climate change impact assessment. *Journal of Hydrology* 153: 265-290.

Wilks, D.S. (1989) Conditioning stochastic daily precipitation models on total monthly precipitation. *Water Resources Research* 25: 1429-1439.

Wolski, P. (1999) Application of reservoir modelling to hydrotopes identified by remote sensing. PhD dissertation. Free University of Amsterdam, ITC Publication No. 69. pp. 191.

Woolhiser, D.A., Keefer, T.O., & Redmond, K.T. (1993) Southern Oscillation effects on daily precipitation in the southwestern United States. *Water Resources Research* 29: 1287-1295.

Woolhiser, D.A., & Roldán, J. (1982) Stochastic Daily Precipitation Models 2. A comparison of distribution amounts. *Water Resources Research* 18: 1461-1468.

Woolhiser, D.A., & Pegram, G.S. (1979) Maximum likelihood estimation of Fourier coefficients to describe seasonal variations of parameters in stochastic daily precipitation models. *Journal of applied meteorology* 18: 34-42.

Xu, C.-Y. (1992) Monthly water balance models in different climatic regions. PhD dissertation. Vrije Universiteit Brussel, Laboratory of Hydrology and Interuniversity Postgraduate Programme in Hydrology. pp. 222.

Zeng, N., Shuttleworth, J.W., & Gash, J.H.C. (2000) Influence of temporal variability of rainfall on interception loss. Part 1. Point analysis. *Journal of Hydrology* 228: 228-241.

Zheng, X., & Eltahir, E.A.B. (1998) A soil moisture-rainfall feedback mechanism 2. Numerical experiments. *Water Resources Research* 34: 777-785.

Zucchini, W., & Adamson, P.T. (1984) The occurrence and severity of droughts in South Africa. WRC report 91/1/84, Dept. of Civil Engineering, University of Stellenbosch and Dept. of Water Affairs, Pretoria, South Africa, pp. 198.

Zucchini, W., P. Adamson, & McNeill, L. (1992) Model of Southern African Rainfall. *South African Journal of Science* 88: 103-109.

References

Acknowledgements

Firstly, I thank very much my promoter, Prof.dr.ir. Hubert Savenije. I am happy that he asked this student whom he met on the river Incomati in Mozambique to come and work with him. I thank him for his endless confidence, for the inspiring discussions and for his attitude that innovative research does not follow a recipe.

I am also very grateful to the Department of Water Resources in Zimbabwe and all its employees. You have shown what Zimbabwean hospitality is all about and through you I learned a lot on real life water resources management. *Tatenda! Ateonana!*

I thank all other PhD researchers who were 'under the umbrella' of the Sharing Scarce Resources Zambezi project. While we had very different research objectives and disciplines, I have learned a lot from you all and enjoyed your company. Your co-operation made my task as a co-ordinator a pleasant task. Ringsom Chitsiko, Piotr Wolski, Denis Rugege, Joylyn Ndoro, Bekithemba Gumbo, Nyasha Lawrence Nyagwambo and Kees de Bie, thank you. In this respect I also thank Prof.dr. Allard Meijerink (ITC), Prof.dr.ir. Paul Driessen (WU/ITC) and dr.ir. Pieter van der Zaag (IHE). I thank the Ministry of Foreign Affairs of the Netherlands, who funded the research project Sharing Scarce Resources Zambezi.

I thank the Department of Meteorological Services of Zimbabwe for making available the data that are the basis of this dissertation. I thank many researchers of this Department, of the University of Zimbabwe, of the Department of Agricultural Technical and Extension Services (Agritex) and of other institutes in Zimbabwe and Zambia for useful discussions. A question by Kwinisha Bwanali, at the time student at the University, triggered the derivation of equations for confidence limits.

I am very grateful for the critical reviews by the reading committee: Prof. Andràs Bárdossy (University of Stuttgart), Prof.dr. Peter Troch (Wageningen University) and Prof.dr. Mike Hall (IHE). Ir. Paul Torfs (Wageningen University) gave valuable comments for Chapter 3, dr.ir. Pieter de Laat for Chapter 10 and Appendix C. Dr. Vincent Guinot, dr. Yanxiou Zhou, dr.ir. Jan Vermaat, ir. Slavco Velickov (all IHE), dr. Ed Veling (TU Delft) and ir. Jaap Karelse were helpful in thinking along on mathematical questions.

I also am very grateful to all colleagues at IHE, including those of Van Hecke Catering. Thanks for your support and company. I thank all MSc researchers whose research I have somehow been involved in, for sharing with me their perceptions on water resources management. In chronological order: Steve Shoniwa, Abebe Belachew Belatu, Vincent Gbedzi, Hodson Makurira, Albert Rockson, Albert Mare, Nyasha Lawrence Nyagwambo (again), José Matola, Elisha Nhidza, Ngoni Mudege, Samhan Samhan, Ismael Seyam, Adane Abebe Awass. Resource Analysis, my current employer, is thanked for flexibility.

I am very grateful to Eva Keller and Jane Madembo who shared with me their respective homes in Harare, while the family Luxemburg's house was a hospitable Dutch harbour over there. Ms Annemarie Weitzel and Stephen Ward corrected my 'double Dutch'. My parents get an applause for the past 32 years. Family and friends, wherever you are in the world, thank you very much.

About the author

Marieke de Groen was born on November 4th 1969 in Nijmegen, The Netherlands. In 1988 she obtained her secondary school degree at the Bouwens van der Boye College in Helden and started her studies Civil Engineering at the Delft University of Technology. She chose to specialize in hydrology and water management. Her first practical experience in this field was in 1993, when for three months she was a guest at the National Directorate for Water Affairs in Mozambique studying salt intrusion in the estuary of the river Incomati. In 1994, at Delft Hydraulics, she conducted her graduation research on the propagation of pollution clouds in the river Rhine. The research involved two-dimensional modelling of the river Rhine downstream of its confluence with the river Mosel and was supervised by Prof. Matthijs de Vries from river engineering. In February 1995 she received her MSc with distinction and was additionally awarded a certificate for half a year extra-curricular exams.

Soon after, she started at the International Institute for Infrastuctural, Hydraulic and Environmental Engineering (IHE) in Delft as the co-ordinator of the research project 'Sharing Scarce Resources Zambezi'. This project was a co-operation between IHE Delft, Wageningen University, the International Institute for Aerospace Survey and Earth Sciences in Enschede (ITC) and Management School Maastricht. They joined forces to conduct PhD and MSc research in Zimbabwe and Zambia on topics related to natural resources management. As a co-ordinator she was responsible for logistics, finances, external relations, progress reports and the organisation of seminars. Marieke's own research, under supervision of Prof. Hubert Savenije, related to the effects of land cover changes on the atmospheric water balance. From September 1995 till May 1997, she and the other project members were guests at the Department of Water Resources of Zimbabwe. On her return she became a lecturer in water resources management at IHE, with a specialisation in the modelling of river basins for strategic planning. In 1999, the PhD research topic evolved its focus on the modelling of interception and transpiration at monthly time steps. To devote more time to the research, she changed to a part-time job in June 2000.

Since March 2001 she worked for the research and consultancy company Resource Analysis, which specialised in policy advice for natural resources management. As consultant in water management, her work is system analysis and the facilitation of actors and stakeholders in policy development, from local scale (water boards) to international scale. Although the work mainly relates to The Netherlands and its neighbouring countries, her first assignment brought her back to the river Incomati in Southern Africa.

www.ihe.nl
www.resource.nl

Appendix A

Use of Transition Matrices for Two State Markov Processes

Matrix computation helps when deriving properties of Markov processes. In my opinion, for two state Markov processes the derivation of equations is less transparent when matrix computation is used. However, for Markov processes with more than two states and for higher-order autoregressive schemes, the use of matrix computation becomes inevitable. The principle of using matrices is described below, based on explanations in mathematical textbooks (Cox & Miller, 1965; Hoel, Port & Stone, 1972; Grimmett & Stirzaker, 2001). In Cox & Miller (1965) the Markov processes applied to rainfall occurrence, as determined by Gabriel & Neumann (1962), are used as examples. Haan et al. (1976) also used matrix computation in the context of rainfall occurrence.

This appendix shows how discrete autoregressive processes are dealt with, in the way they are used in this dissertation. Any x state discrete autoregressive process of the k-th order can be regarded as two state processes of the x^kth-order, as used by Foufoula-Georgiou & Georgiou (1987). Here the example of a two states Markov process is shown. In stochastic daily rainfall models a third state is optional and usually refers to the occurrence of trace rainfall. For more or fewer states rows and columns can be added, respectively deleted.

	states	actual day 0 (dry)	1 (wet)
preceding day	0	p_{00}	p_{01}
	1	p_{10}	p_{11}

Logically, the actual day can only be in one of the two states. Therefore the totals of the probabilities in the rows should be 1 (e.g. $p_{00} + p_{01} = 1$).

The conditional probabilities can be placed in a matrix. In the literature such a matrix is usually referred to as a **transition matrix**:

$$\mathbf{P} = \begin{bmatrix} p_{00} & p_{01} \\ p_{10} & p_{11} \end{bmatrix} \qquad \text{Eq. A.1}$$

The transition matrix is a stochastic matrix, the defining property of a stochastic matrix being that its elements are non-negative and that all its row totals are unity.

In addition to matrix $\mathbf{P}$ a row vector $\mathbf{p}$ is defined, which denotes the probabilities of finding the system in a certain state at time n:

$$\mathbf{p}_0^n = \begin{bmatrix} p_0^n & p_1^n \end{bmatrix} \qquad \text{Eq. A.2}$$

The Markov property assumes that the occurrence of a certain state only depends on the state in the previous time step.

Therefore:

$$\mathbf{p}^{(n)} = \mathbf{p}^{(n-1)}\mathbf{P}$$

Eq. A.3

and on iteration

$$\mathbf{p}^{(n)} = \mathbf{p}^{(n-2)}\mathbf{P}^2 = \cdots = \mathbf{p}^{(0)}\mathbf{P}^n$$

Eq. A.4

where $\mathbf{p}^{(0)}$ is a row vector with initial conditions.

N.B. Matrix multiplication means that

$$\begin{bmatrix} a & b \\ c & d \end{bmatrix}\begin{bmatrix} e & f \\ g & h \end{bmatrix} = [(ae+eg) \quad (cd+fh)]$$

Eq. A.5

For successive time steps the influence of the initial condition becomes smaller. The row vector $\mathbf{p}^{(n)}$ approaches an equilibrium row vector π for large n. Clearly π satisfies

$$\pi = \pi\mathbf{P}$$

Eq. A.6

which means

$$\pi(\mathbf{I}-\mathbf{P}) = 0$$

Eq. A.7

where $\mathbf{I}$ is the unit matrix, in this case with two states $\mathbf{I} = \begin{bmatrix} 1 & 0 \\ 0 & 1 \end{bmatrix}$. The only way for π

not to be a zero matrix is if the determinant $|\mathbf{I}-\mathbf{P}|$ vanishes. Together with the fact that the sum of a row in the transition matrix is 1, this will lead to a solution for π.

For the two state Markov model the solution of π is $\pi = [(1-p) \quad p]$, where $(1-p)$ is the probability of a dry day and p is the probability of a rain-day (p) and the solution

is $p = \dfrac{p_{01}}{1 - p_{11} + p_{01}}$, as in Eq 4.6.

The transition matrix $\mathbf{P}$ has distinct eigenvalues λ, which can be found by solving $|\mathbf{P} - \lambda\mathbf{I}| = 0$. This yields $\lambda_0 = 1$ and $\lambda_1 = p_{11} - p_{01}$. This will simplify the computation of $\mathbf{P}^{(n)}$, because

$$\mathbf{P}^n = \mathbf{Q}\begin{bmatrix} \lambda_0^{\ n} & 0 \\ 0 & \lambda_1^{\ n} \end{bmatrix}\mathbf{Q}^{-1}$$

Eq. A.8

where the columns $\mathbf{q}_1, \mathbf{q}_2, \mathbf{q}_3$ of $\mathbf{Q}$ are solutions of the equations

$$\mathbf{P}\mathbf{q}_i = \lambda_i\mathbf{q}_i$$

Eq. A.9

Subsequently $\mathbf{Q}$ is

$$Q = \begin{bmatrix} 1 & p_{01} \\ 1 & p_{11}-1 \end{bmatrix}$$

Eq. A.10

Using these matrices, many of the probability density functions presented in Chapter 4 can be derived.

The above explanation can be used to derive the equations that have been described in this dissertation for a two state process for higher orders. Also, it is possible to prove the equations that have been quoted from other sources. Matrix computational programmes, such as *Matlab*, are of great assistance in solving the above equations.

Appendix B

Gamma and Weibull Distribution

In the main text reference is made several times to gamma and Weibull probability density functions. In this Appendix they are described. For this description use is made of Benjamin & Cornell (1970), Grimmett & Stirzaker (2001) and Stedinger et al. (1997).

Gamma distribution

For $P_r > 0$ (mm/day); α (mm/day) > 0 and k (-) > 0:

$$f(P_r) = \frac{\left(\frac{1}{\alpha}\right)\left(\frac{P_r}{\alpha}\right)^{k-1} \exp\left[-\left(\frac{P_r}{\alpha}\right)\right]}{\Gamma(k)} \qquad \text{Eq. B.1}$$

$$\Gamma(k) = \int_0^\infty \exp[-u](u)^{k-1}\, du \qquad \text{Eq. B.2}$$

$$F(P_r) = \frac{\Gamma(k, \frac{P_r}{\alpha})}{\Gamma(k)} \qquad \text{Eq. B.3}$$

$$E(P_r) = \alpha k \qquad \text{Eq. B.4}$$

$$\text{Var}(P_r) = \alpha^2 \left\{ \Gamma\left(1 + \frac{2}{k}\right) - \left[\Gamma\left(1 + \frac{1}{k}\right)\right]^2 \right\} \qquad \text{Eq. B.5}$$

The gamma distribution with $k = 1$ is really the exponential distribution. A gamma distribution with $k = 1/2$ is the chi-squared distribution $\chi^2(P_r)$.

Although probability density functions are used here for rain on rain-days they originate from models for random occurrences. If events occur at random times then the gamma k distribution is the time since the last event until the kth arrival of another event. (The probability density function of the number of events in a certain interval then has a Poisson distribution, while the probability density function of inter-arrival times is exponential.)

Weibull distribution

For $P_r > 0$ (mm/day); α (mm/day) > 0, k (-) > 0:

$$f(P_r) = \left(\frac{k}{\alpha}\right)\left(\frac{P_r}{\alpha}\right)^{k-1} \exp\left[-\left(\frac{P_r}{\alpha}\right)^k\right] \qquad \text{Eq. B.6}$$

$$F(P_r) = 1 - \exp\left[-\left(\frac{P_r}{\alpha}\right)^k\right]$$
         Eq. B.7

$$E(P_r) = \alpha\, \Gamma\left(1 + \frac{1}{k}\right)$$
         Eq. B.8

$$\mathrm{Var}(P_r) = \alpha^2 \left\{\Gamma\left(1 + \frac{2}{k}\right) - \left[\Gamma\left(1 + \frac{1}{k}\right)\right]^2\right\}$$
         Eq. B.9

The Weibull distribution with k is 1 is an exponential distribution.

Weibull distributions originate from models of limiting cases. If X has a Weibull distribution then $Y = -\ln(X)$ has a Gumbel distribution. This allows parameter estimation procedures available for the Gumbel distribution to be used for the Weibull distribution (Stedinger et al., 1997).

Appendix C

Equations for Potential Evaporation

The Penman and Penman-Monteith equations are established physical approximations for potential evaporation which have been used in Chapter 10. Their descriptions can be found in most textbooks on hydrology. However, because the symbols are not used consistently in different textbooks and for the sake of completeness, this dissertation also includes them. The following descriptions make extensive use of De Laat (1997).

Penman

The Penman equation for open water is applicable as an estimate for actual evaporation from a large shallow lake (neglecting advection and storage of heat below the surface). It reads:

$$E_o = \frac{C}{L} \frac{s R_N + c_p \rho_a (e_a - e_d)/r_a}{s + \gamma} \qquad \text{Eq. C.1}$$

E_o Penman open water evaporation   (mm.d^{-1})

Variables taken constant:

C	constant to convert units ($C = 86400$)	(mm.d^{-1}/(kg.m^{-2}.s^{-1}))
L	latent heat of vaporisation ($L = 2.45 * 10^6$)	(J.kg^{-1})
c_p	specific heat of dry air at constant pressure ($c_p = 1013$)	(J. kg^{-1}. °C^{-1})
γ	psychrometric constant $c_p p_a/(\varepsilon L) \approx 0.00066\, p_a$	(kPa.°C^{-1})
	(γ = for altitude Harare 0.056)	
	where	
p_a	is air pressure	(kPa)
ε	ratio of molecular masses of water vapour and dry air ($\varepsilon = 0.622$)	(-)

Variables dependent on surface and radiation (which is also related to temperature):

r_a aerodynamic resistance for open water $r_a = 245/(0.54 U_2 + 0.5)$ (s.m^{-1})
       where
U_2 is wind speed, 24 hour mean at 2 m above surface   (m/s)
R_N net radiation at the earth's surface   (W.m^{-2})
       which is the non-reflected part of the short-wave radiation minus the net outgoing long-wave radiaton: $R_N = (1-r) R_S - R_{nL}$.
       where
r is the factor of reflection, commonly called albedo (the Penman equation itself is for free water, where $r = 0.06$; for Penman-Monteith for grass $r = 0.22$-0.25, bare soil $r = 0.10$-0.30)
R_S short-wave solar radiation, which can be measured with a solarimeter or estimated $R_S = (a + bn/N)R_A$, where n is the actual hours of sunshine, N the possible hours and R_A the short-wave radiation received at the outer limits of the atmosphere (function of season and latitude); the parameters a and b are empirically derived (Torrance, 1981; for Zimbabwe $a = 0.29$ cos(latitude), $b = 0.52$)
R_{nL} net outgoing long-wave radiation: (W/m^2)

$$R_{nL} = \sigma \frac{(273 + T_{min})^4 + (273 + T_{max})^4}{2} \left(0.34 - 0.139\sqrt{e_d}\right)\left(0.1 + 0.9\frac{n}{N}\right) \qquad \text{Eq. C.2}$$

where T_{min} and T_{max} are respectively daily minimum and maximum temperature (°C) and σ is Stefan-Boltzmann constant ($\sigma = 5.6745 \times 10^{-8}$ W.m^{-2}.K^{-4})

Variables dependent on temperature and water vapour pressure:

e_a saturation vapour pressure $0.6108 \times \exp(17.27\, T_a / (T_a + 237.3))$ (kPa)
         where

T_a is actual temperature (mean between maximum and minimum daily temperature)   (°C)

e_d actual vapour pressure: $e_a\, RH/100$ (kPa)
         where

RH is relative humidity   (%)

s slope of saturation vapour pressure-temperature curve:
         $de_a/(dT_a) = 4098 e_a/(T_a + 237.3)^2$ (kPa. °C^{-1})

ρ_a atmospheric density: $p_a/(0.287 \times (T_a + 275))$ (kg.m^{-3})

Penman-Monteith equation

The Penman-Monteith equation for evaporation from a wet crop is almost exactly the same, but values for the net radiation (albedo r) and the aerodynamic resistance r_a are different. Thus:

$$I_{pot} = \frac{C}{L} \frac{sR_n + c_p \rho_a (e_a - e_d)/r_a}{s + \gamma} \qquad \text{Eq. C.3}$$

Potential transpiration is lower than potential evaporation from a wet surface because of crop resistance. The Penman-Monteith equation reads:

$$T_{pot} = \frac{s + \gamma}{s + \gamma(1 + r_c/r_a)} I_{pot} \qquad \text{Eq. C.4}$$

where
r_c is crop resistance   (s.m^{-1})

With more water stress, the crop resistance r_c increases. However, in this dissertation the Penman-Monteith equation is used to determine non-water-constrained potential transpiration. The actual soil moisture content is then accounted for by making the actual transpiration a function of the potential transpiration and soil moisture content. Therefore for r_c minimum values are used which are characteristic for a certain land cover.

As is shown in this dissertation (Chapter 10), measurements of pan evaporation, relative humidity, and minimum and maximum temperature in the day are considerably different on dry and on rain-days. This affects the potential evaporation, according to Penman. In the Penman-Monteith equation for transpiration, potential evaporation from a wet surface is therefore referred to as $I_{pot,dry}$ for dry days and $I_{pot,r}$ for rain-days. Similarly, two forms of potential transpiration are distinguished: $T_{pot,dry}$ and $T_{pot,r}$. In practice, $I_{pot,dry}$ only has meaning in the case of sprinkler irrigation.

For rain-days the potential transpiration is limited by the actual interception:

$$T_{pot,w} = \frac{s + \gamma}{s + \gamma(1 + r_c / r_a)}(I_{pot,r} - I)$$

Eq. C.5

FAO Modified Penman

The FAO Modified Penman equation is the most widely used method for computation of reference crop evaporation. The reference crop is an extensive surface of an 8 - 15 cm tall green grass cover of uniform height, completely shading the ground and not short of water. Any other crop should have a potential transpiration proportional to this crop, for which crop factors are available in the literature (Doorenbos & Pruitt, 1977).

$$T_{pot.ref,pen} = c\left[\frac{s}{s + \gamma}\frac{C}{L}R_N + \frac{\gamma}{s + \gamma} * 2.7 * f(U) * (e_a - e_d)\right]$$

Eq. C.6

where

c is an adjustment factor. For moderate winds, a maximum relative humidity at night of about 70% and a day-night wind ratio of 1.5-2, the adjustment factor, c, equals 1; although daily mean relative humidity is regularly higher in Grasslands, for computations c has simply been assumed 1

$f(U)$ wind function; $f(U) = 1 + 0.864\,U_2$ (mm/day*kPa^{-1})

The wind function replaces the aerodynamic resistance r_a in the Penman equation. Additionally for R_N (see above) the albedo $r = 0.25$.

FAO Penman-Monteith

In 1990 FAO decided that the future reference transpiration should be the Penman-Monteith potential transpiration, using a hypothetical crop with fixed albedo r and crop resistance r_c (Smith, 1991)

r $= 0.23$

r_c $= 20$ s/m

and an aerodynamic resistance that depends on the average wind speed at a height of 2 m.

r_a $= 208/U_2$ s/m

The Penman-Monteith transpiration for this hypothetical crop closely resembles transpiration from an extensive green grass cover of uniform height, actively growing, completely shading the ground and not short of water. Because adjusted cropping factors are not yet available, this method is not yet widely used. However, in the main text of this dissertation it has already been used for comparison with pan evaporation and is referred to as $T_{pot,\ ref.\ penmon}$.

Milton Keynes UK
Ingram Content Group UK Ltd.
UKHW050259161024
449569UK00043B/1484